WOODY ORNAMENTALS for the PRAIRIES

Revised Edition

Copyright © 1995

University of Alberta

Faculty of Extension

Edmonton, Alberta T6G 2T4

All rights reserved. No part of this publication may be reproduced, stored in a retrieval system, or transmitted in any form or by any means whatsoever without prior permission of the copyright owners.

Canadian Cataloguing in Publication Data

Knowles, Hugh.

    Woody ornamentals for the Prairies

    ISBN 1-55091-025-6

1. Ornamental woody plants – Prairie Provinces. I. University of Alberta. Faculty of Extension. II. Title.
SB435.6.C32P73 1995   635.9'7   C95-910241-8

Previous edition published under ISBN 0-88864-869-3 (1989)

Production Team

Managing Editor	Thom Shaw
Graphic Artist	Melanie Eastley
Page Composition	Lu Ziola
Copy Editor	Lois Hameister
Technical Reviewers	Brian Baldwin
	Brendan Casement
Cover Photographs	Hugh Knowles
	Brian Baldwin
	J.O. Hrapko

Contents

Preface iv
How to Use This Book v
Acknowledgements vi

Chapter 1 Plant Selection

Landscape Values 1
    Enframement 2
    Accentuation 3
    Skyline Articulation 3
    Background 3
    Screening 3
    Enclosure 3
    Climate Control 3

Landscape Concerns 4
    Functional Considerations 4
    Aesthetic Considerations 5

Chapter 2 Planting and Maintenance

Planting Procedures 7
    Nursery Stock 7
    Planting Depth 8

Soils and Nutrition 8
    Soil Texture and Structure 8
    Plant Nutrition 9

Pruning Practices 10
    Pruning at Planting Time 10
    Pruning Cuts 10

Annual Pruning 12
    Spring-Blossoming Plants 12
    Summer-Blossoming Plants 12

Pest Management 13

References 14

Chapter 3 Woody Plant Problems

The Prairie Region 15
    Inter-Regional Climates 15
    Intra-Regional Climates 15

Problem Types 16
    Low Temperature Problems 16

Plant Diseases 18
    Fungal Diseases 18
    Bacterial Diseases 19
    Viral Diseases 19
    Mycoplasmal Diseases 19

Insect Problems 20
    Chewing Insects 20
    Sucking Insects 20

References 21

Chapter 4 Planting Design

Planning and Design 22
    Investigation 22
    Plant Selection 22
    Size, Form, Color, Texture and Value 25
    Principles of Composition 26

References 29

Plant Descriptions 31

Reference Charts 171
    Coniferous Shrubs 172
    Coniferous Trees 178
    Deciduous Shrubs 180
    Deciduous Trees 196
    Groundcover Plants 204
    Vines and Climbers 206

Glossary 207

Cross-Referencing Index 210

Woody Ornamentals

Preface

Woody Ornamentals for the Prairies is one of the University of Alberta's Home Gardening Series. It aims to provide the user with the sort of information needed when selecting plants not simply for a particular environment or for a particular function, but more importantly for a particular function within a particular environment. In addition, the book contains four chapters that address subjects of importance to all who would use trees, shrubs and groundcovers in the articulation of garden space.

Because woody plants are the major elements of the landscape, the new gardener will soon learn that it is important to be familiar with their various shortcomings as well as their capabilities. Chapter One, **Plant Selection,** provides an overview of the important things to be considered when selecting a plant for a particular purpose.

No book on woody landscape plants would be complete without some discussion of **Planting and Maintenance.** This chapter begins with something on planting principles, proceeds to the subject of soils and nutrition, and finally to pruning practices and pest management, two topics in which the principles have been subject to re-appraisal and revision in recent years.

The chapter **Woody Plant Problems** relates in part to the more common plant stresses and disorders associated with the general climate of the region. The more common insect and disease problems affecting woody ornamental plants are also examined.

The final chapter, **Planting Design,** addresses both the basic aesthetic and ecological decisions to be made when using woody plants to enhance a garden space. The ecological side of planting design is, in part, site specific and relies very much on experience. In dealing with the aesthetic aspect, however, the reader is referred to the principles of visual design that pervade everything from Fine Arts to Architecture.

The sections **Reference Charts** and **Plant Descriptions** are not restricted to just trees and shrubs. Vines and groundcovers are also included and at least some of the latter are non-woody. The charts, color plates, and silhouettes in these sections will help the user to make the appropriate choice (more about this in the section "How to Use This Book").

Botanical and cultivar names used in this edition follow the usage of *Hortus III*. The nomenclatural rules for common names are to be found in *The Spelling of Common Names* by Robert A. Hamilton, Research Branch, Canada Agriculture, Ottawa.

A **Glossary** of botanical and horticultural terms has been included to clarify the meaning of unfamiliar words used in the plant descriptions.

One of the problems that an author encounters when writing on this subject involves decisions on what to include and what to leave out. Invariably plants will be included that are not commonly available. There is no question that the user will find plants described and recommended in this book and yet not find them available from local sources. However, if recommendations were to be restricted to only those plants that were readily available, there would be no incentive for nursery people and garden centre operators to stock anything new.

It has been mentioned on more than one occasion that the mesoclimate can have a major influence on the response of many woody plants to winter conditions. Because of this, we must expect environments within the influence of major cities to be more favorable for winter survival. For this reason, I would not hesitate to suggest other sources of supply to most city dwellers when a recommended plant is not available locally. However, I would be less than responsible if I did not say that the safest source of hardy woody plants still reside with the local grower.

Hugh Knowles, PAg, FCSLA

March 1995

How to Use This Book

Woody Ornamentals for the Prairies was written primarily to help people in the choice and selection of woody plants for use in the landscape.

One of the major problems encountered in putting together a book of this sort is how to overcome the difficulties associated with botanical names with which many of its users are unfamiliar. In this, the revised edition, we hope that botanical names will no longer be a problem, not that we have abandoned botanical nomenclature and the international rules for its usage, but rather because we have added an indexing system that cross-references the common name of a plant to its generic equivalent. This, we hope, will make the book much easier to use.

The book has three major sections. The first consists of four chapters in which are discussed the criteria for the selection, use and maintenance of woody plants. The publication is not one of those how-to books, so the reader will find that this first section simply provides only that sort of fundamental information to which all gardeners in the prairie provinces should relate before undertaking any landscape project with woody plants.

The subject matter of the other two sections is totally descriptive. If you are inexperienced and you are looking for a plant with a particular form, size, and color to fit a special space, we suggest that you try making your selection by first referring to the appropriate chart. There are six of these to choose from: deciduous and coniferous trees, deciduous and coniferous shrubs, vines, and groundcovers.

The characteristics referred to in the charts were chosen to help you make a first selection. Page numbers where further information can be found are included.

Should you have no need to use the charts, the Cross-Referencing Index will lead you to the page number on which the generic name of the plant can be found. Detailed description will be found in this general area.

If you have a botanical background, use of the index will be unnecessary because all plants described are listed alphabetically by generic name.

A glossary of botanical and horticultural terms is also included.

Acknowledgements

It is always a pleasure to acknowledge the support and encouragement received from friends and colleagues when a project of this sort is finally put to bed.

The search for suitable photographic material has always been of major concern to me. I never ever seem to have had a broad enough selection in my own collection due to one catastrophe or another, and I am grateful to those who have let me have what in some cases has amounted to free access to theirs.

I am much obliged for the help received from the University of Saskatchewan, Saskatoon; from Alberta Agriculture; from the Northern Forestry Centre, the U of A Devonian Botanic Garden, and the Northern Alberta Institute of Technology in Edmonton. I also have special thanks for contributors Brian Baldwin from Saskatoon; Dieter Martin, Langham, Saskatchewan; Campbell Davidson from Morden, Manitoba; Brendan Casement, Julie Hrapko, Ieuen Evans, Bruce Williams, Ed Toop, Ken Mallett, Charlotte Smith, and Bruce Dancik from Edmonton; John Davidson from Beaverlodge, Alberta; and Casey Van Vloten from Pitt Meadows, B.C., all of whom have offered slides from their private collections.

Again, I am pleased to extend my thanks and show my admiration for the fine work of the production team in putting the book together and for the quality, effort and exceptional job done by Brian Baldwin of Saskatoon with the technical review of this edition. It has been a pleasure to work and collaborate with all members of such a talented group.

Finally, my thanks to Jessie, my wife, to whom the book is dedicated.

Hugh Knowles, PAg, FCSLA

March 1995

Chapter 1
Plant Selection

If woody plants could be chosen for landscape purposes solely on the basis of use and appearance, selection would be a fairly simple task. Unfortunately, we must also know the specific environment to which each plant is best adapted and be aware of undesirable characteristics and the pests to which it is subjected. Failure to take any one of these things into account can only lead to disappointment and frustration.

Because each plant species has a preferred habitat, many problems can be avoided by knowing their growth requirements beforehand. For instance, gardeners should be aware that pines normally grow in dry, sunny environments and yet many people will plant mugo pine on the north side of a building where soil moisture is highest and sunlight is in shortest supply. Mugo pine seldom retains its compact form for more than a few years in such an environment. On the other hand, the broad-leaved groundcover Japanese spurge does beautifully in the deepest shade but soon acquires an unhealthy paleness when grown in full sun.

Some plants are susceptible to specific diseases. For example, Dutch elm disease literally wiped out the American elm in the eastern part of the continent. This disease has now been reported as far west as Saskatchewan and Montana and it is expected that it will be only a matter of time before it reaches Alberta. Should elm trees still be planted? Similar questions might be asked about flowering crab apples, mountain-ash and Ussurian pear. Each of these plants is susceptible to the bacterial disease fire-blight which can also be absolutely devastating because of its very infectious nature.

Insect problems, too, can be serious in many instances. Recently, birch trees have lost some of their attraction due to the ravages of an insect pest. Annual infestations of birch leaf miner have been so bad in many parts of the country that they have caused much of the foliage of these trees to drop prematurely.

And then, of course, there are those undesirable growth habits that must also be considered. It has been stated many times that when it comes to using trees there is no "free lunch." Most trees, it seems, have something in their make-up that can create problems for the gardener. There are the female poplars and their "cotton", Ussurian pear and its fruit, mayday tree and its suckers; yes, all trees have bad habits of one sort or another. Fortunately, these are much more tolerable in some cases than they are in others.

Landscape Values

Woody plants are the major structural elements of the landscape. Trees are the strongest of these and are primarily selected for purposes of enframement, accentuation, skyline articulation, background, enclosure, screening, and climate control.

It is not difficult to find trees being used for any one or all of these purposes in the larger landscape, but in the smaller landscape situations, such as residential gardens, the primary use of trees is quite naturally limited.

Figure 1. Enframement

Woody Ornamentals

Enframement

In all landscapes, large and small, trees are used to enframe views. Enframement is a common technique used by the nature photographer to relate some distant subject to the point from where the picture was taken. This is frequently accomplished by including some foreground element of vegetation at the top and side of the picture. For this type of enframing element, nothing can quite match the value of a high-headed tree with a large canopy. We live in a region of "big sky," and it is often necessary to reduce its impact in order to isolate and identify the important parts of a view or garden. This is also one of the reasons why the major "people space" in a small garden seems like a pleasant place to be if it is located adjacent to or beneath the canopy of a high-headed tree. This is also one of the reasons why a garden space without trees seems bare, exposed, and rather uncomfortable.

Figure 2. Problems and Shortcomings

	Diseases						Insects								Miscellaneous							
	Black-knot	Canker	Crown gall	Fire-blight	Rust	Shoestring fungus	Aphids	Beetles	Borers	Cankerworms	Leaf miners	Pear slugs	Mites	Tent caterpillars	Climate problems	Deciduous branches	Poisonous fruit	Poisonous leaves	Fruit	Shallow roots	Suckering	Tender flower buds
Acer negundo							•	•	•										•			
Aegopodium																					•	
Aesculus																	•					
Amelanchier		•			•																	
Betula									•		•				•	•				•		
Cotoneaster				•								•										
Crataegus				•	•					•												
Daphne																	•	•				
Forsythia															•							•
Fraxinus							•	•		•										•		
Hippophae																					•	
Juniperus				•									•									
Lonicera							•															
Malus				•																		
Picea							•						•									
Populus		•		•																•	•	•
Prunus fruticosa																						•
Prunus maackii						•									•					•		
Prunus padus var. commutata	•									•				•						•	•	
Pyrus ussuriensis				•																•		
Quercus																				•		
Rhus glabra																				•		
Rosa rubrifolia			•																			
Salix																•						
Sorbaria															•							
Sorbus				•								•										
Syringa vulgaris											•										•	
Tilia														•					•			
Ulmus							•			•						•					•	

2

Woody Ornamentals

Figure 3. Accentuation

Figure 4. Skyline Articulation

Accentuation

In the design of the large landscape, the need to draw attention to a particular spot or to a particular feature is frequently encountered. Trees with unusual form, color or both, are very useful for this purpose. The tall columnar tower poplar, the golden-barked willow, the silver-leaved Russian olive, and the red-foliaged Shubert chokecherry are some of the better ones. In some cases a single plant will be all that is required, but in others it may take several plants to achieve the impact required. Plants with seasonal color characteristics such as those displayed by flowering crab apples, plums, cherries, and lilacs are also good accent materials, as are plants with good fall color. When a tree is found that combines several distinctive characteristics, its value as an accent plant is greatly increased.

Skyline Articulation

Within the enclosed space of a small garden it is very easy for enclosing elements like fences and walls to become dominant. Very often the top of a fence will lie at or near eye level and present a very strong horizontal line. To relieve this, trees can be used in visual conjunction with the fence to create an undulating skyline and mask or soften the elements of the adjoining neighborhood.

Background

In all landscapes, background is very important to the display of garden features; consequently, the elements of background must be appropriate for the feature being displayed. A piece of light-colored sculpture will read much more strongly in front of a dense background of evergreens, but may be lost completely against a varied background of multicolored shrubs.

Screening

Trees are often used for screening purposes, but some suit the function better than others. Choice is dependent to a large degree on the size of the landscape. In small landscapes the low-headed, narrowly upright tree is much more satisfactory, whereas in the large landscape, where space is not so limiting, both high- and low-headed trees can be effectively used for screening purposes.

Enclosure

As enclosing elements, trees have great value on the large property but have little or no value on a small property. Trees as elements of enclosure take up far too much space on the small property. They are no substitute for fences and walls which can provide enclosure and still be no more than a few centimetres wide.

Figure 5. Background

Climate Control

Improper siting of houses sometimes results in problems which call for expensive solutions to achieve the level of comfort expected within the home. Sometimes the bedrooms are located in the part of the house that cools off last in the evening. Perhaps the view from the living room is oriented to the south and the sun is fading the furniture. In each case, a well placed tree or trees will do much to substitute for expensive equipment. It is also possible to have trees that will give shade during the summer but permit light to enter when it is more highly valued during the winter.

Landscape Concerns

Functional Considerations

Deciduous Trees

In small landscapes, high-headed trees are selected for the more heavily used areas simply because they are more compatible with human activity. People can move around beneath such trees without having to worry about losing an eye to a low-hanging branch. Except for the space occupied by the trunk, the surface below the tree remains usable.

While the high-headed form is the more practical tree type where gardens are for people, it is important to select trees that are not going to drop a great deal of unwanted trash on the area below. Such might well be the case if a high-headed flowering tree or a tree with an inherent insect problem was selected. It is not altogether uncommon to find people still using such things as Manitoba maple to provide canopy over patio areas only to find that it attracts aphids. The quick-growing mayday tree has also been widely used for the same purpose and people have learned, to their dismay, that it produces an abundance of fruit which begins dropping in early August and is tracked around on shoes.

Because of their low branching habits, low-headed trees must be used carefully in the small garden. Nevertheless, they can be very useful on the edges of garden space where they may be used to fulfill a number of functions. They are good elements for screening, for providing and creating an interesting skyline, and for defining space. A few low-headed species, however, must be considered as exceptions to this. The cutleaf weeping birch is one, simply because of the great amount of space it occupies at ground level. Because of its size and stature, this tree is much more satisfactory when placed on a large open lawn rather than within the confines of a small garden space.

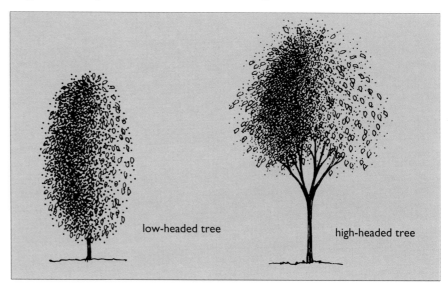

Figure 6. Major Elements: Low- and High-Headed Trees

Conifers

Coniferous trees are highly valued elements for northern landscapes. Generally, they are very strong subjects because of their form, color, and density.

Small numbers of conifers usually go a long way in the landscape. There is no question of their value in the winter, but as with other trees, some discretion must be used to optimize their use. A rather interesting contradiction exists with respect to their usage. If too many are used, the effect, at least during the summer, can become very sombre. On the other hand, if a few well-chosen and well-placed conifers are used in a predominantly deciduous landscape, they can provide a good deal of interest.

Many people are unaware that some high-headed conifers make acceptable shade trees. Scots pine, with its attractive and somewhat open canopy, is one worth considering. It should also be pointed out that only when canopy and shade are valued twelve months of the year should such use be considered. In situations where summer shade and winter sunlight are more highly valued, a high-headed deciduous tree is more suitable.

Shrubs

While trees are considered one of the important infrastructural elements of the landscape, shrubs too are important. Shrubs might be considered to be of even greater significance than trees in some small landscapes, although they are of secondary importance in large-scale situations.

One of the major contributions of shrubs to the landscape is the variety of form and stature they provide. In the prairie region, people have far fewer small shrubs from which to choose than do people from southern Ontario and British Columbia; still, there should be no excuse for choosing things that will ultimately grow to be out of scale.

In the large landscape, taller shrubs, in the order of 1 to 3 m high, are of much greater significance than are shrubs of less than 0.5 m. In such landscapes, large shrubs are commonly used as edge material to face down a tree mass when, and if, called for by the design.

While trees are frequently used as single specimens in the landscape, this is seldom, if ever, the case with shrubs. In landscape design, shrubs are used in either small groups or large masses. If the landscape is small-scale and residential, the shrubs used will be mainly small-statured, with just the occasional tree or larger shrub introduced for accent purposes. On the other hand, if the landscape is large-scale, then larger shrubs will predominate and trees will also assume a more significant role.

In larger landscapes, the elements of accentuation are not generally assigned to any designated plant type. Small groups of shrubs or trees within the overall planting are often used and frequently repeated to provide the accent necessary. Small flowering trees can also be effectively used as accents when placed just in front of some of the major plant masses.

Groundcovers

Most landscape authorities would never consider a planting plan finished until some thought had been given to the use and location of groundcovers.

The landscape value of groundcovers lies partly in their ability to provide a horizontal link between shrub groups and partly in providing cover beneath trees, or in those places where the use of grass or other plants would not be practical. In addition, a good mass of groundcover will do a lot to enrich the character of the ground plane. As an added bonus, they do not need weekly mowing!

In the larger landscapes, many plants used as groundcovers would be considered shrubs in the small residential landscape. For example, it is not uncommon in the large landscape to find many of the dwarf junipers fulfilling the groundcover role. Also, many of the smaller dwarf deciduous shrubs, like gold flame spirea, dwarf euonymus and dwarf viburnum, may assume groundcover status in these situations.

In the small residential landscape, groundcovers used are almost always low, ground-hugging plants. Types vary, with broad-leaved evergreens, deciduous, herbaceous, and coniferous species being common. Some have leaf textures as fine as conifer needles; others have textures almost as coarse as rhubarb. Some groundcovers blossom, and some have foliage color other than green. Hence, by manipulation of color and texture between adjoining materials, it is possible to achieve a good deal of interest and richness in the process.

Aesthetic Considerations

While woody plants are primarily selected on the basis of what they can do physically for the space we call the garden, there are important aesthetic considerations to be made as well. These are concerned with plant color, form, and texture – things that can enrich the quality of the space.

Figure 7. A Checklist of Small Shrubs

Conifers	Broad-Leaved Plants
Dwarf Arborvitae	Golden Broom
Dwarf Balsam Fir	Purple Broom
Dwarf Spruces	'Vancouver Gold' Broom
Dwarf Pines	Alder-Leaved Buckthorn
Horizontal Junipers	Russet Buffaloberry
Pfitzer Junipers	Dwarf Winged Burningbush
Savin Junipers	Warty-bark Burningbush
	Roundleaf Cotoneaster
	Pygmy Caragana
	Chinese Bush Cherry
	Shrubby Cinquefoil
	Cliff Green
	Creeping Cotoneaster
	February Daphne
	Rose Daphne
	'Isanti' Dogwood
	'Kelsey' Dogwood
	Dyer's Greenweed
	Oregon-Grape
	Forsythia
	Compact Bush-cranberry
	Dwarf American Bush-cranberry
	Dwarf European Bush-cranberry
	Albert Thorn Honeysuckle
	'Miniglobe' Honeysuckle
	Snow Hills Hydrangea
	Leadplant
	Oregon-Boxwood
	Rest Harrow
	Rose (*Rosa arkansana* hybrids)
	Eastern Sandcherry
	Lemonade Sumac
	Three-Lobed Spirea
	Woadwaxen
	Blue Fox Willow
	Dwarf Pink Spiraeas

Plant Color

Plant color comes from a variety of sources. Some is fleeting, like that displayed annually by lilacs, viburnums, and flowering crab apples in the spring; that which we derive from maples, dogwoods, aspens and pin cherries in the autumn; and from dogwoods, willows and birches in the winter. Some plant color is a more permanent summer feature. Russian olive, Shubert choke cherry, and variegated dogwood can provide color throughout the summer. Both types of color are important landscape articulators; however, the former is the more subtle and dynamic of the two. Also, the static color expression of the latter when overdone, can look very artificial. However, if this color type is used only for accent purposes, and only when and where needed, there is little danger of this happening.

When dealing with color, people should be aware that, while the most common color of woody plants is green, there is a great variety of greens. There is blue-green, grey-green, yellow-green, red-green, purple-green, and black-green. Such variation provides the opportunity for a very subtle use of color in planting design. Like sunlight and shade, the various greens in the hands of a skillful designer can be used to enhance the three-dimensional qualities of plant groups in the landscape.

Season of bloom is closely related to color expression and yet it goes far beyond. When a designer is knowledgeable with respect to this factor, it is possible to achieve a much longer blooming season from woody plants. It is possible, even in the small garden, to have a continuous blossoming season with woody plants of from six to eight weeks by careful selection of materials.

Autumn coloration can also be exploited by gardeners. By knowing a plant's color season as well as the color produced, it is possible to greatly extend and expand the landscape contribution to be made by woody ornamentals.

One final aspect of planting design that most people seem to overlook is that of the winter landscape. Because winters are long in the prairie provinces, anything that will brighten the scene at this time of year can be of real value. Woody plants with brightly-colored bark, persistent fruit, and attractive form have much to contribute.

Form

Form is one characteristic of landscape plants that can strongly influence the effect of the garden on the user. For example, suppose a garden space had room enough for five trees and you had the choice of using either five columnar trees without canopies or five high-headed trees with fairly extensive canopies. Now visualize the effect that each of these two forms would have on the viewer. In the first, the space would be so dominated by the verticality of the trees that little, if any, of the viewer's attention would tend to focus on the other elements of the garden. In the second, the tree canopy would tend to enclose things; consequently, the viewer would pay more attention to those elements that were at and below eye level.

Form can also have an effect on the "hard" or manufactured elements of the landscape. Plants with rounded silhouettes, for example, tend to anchor things and for this reason are frequently used at corners of buildings where a vertical building face meets the horizontal ground plane. Upright-spreading tree forms, particularly those that are multi-stemmed, are very useful for breaking up large blank wall spaces. When something is needed to reduce the apparent width of some vertical element, nothing works better than a couple of columnar plants. Strategically placed, they attract and direct the eye upward, rather than letting it dwell on the horizontal dimension of the offending element. Regardless of the problem, appropriate plant forms can be found to complement most architectural or structural features when they occur in the landscape.

Plant Texture

Plant texture is largely dependent upon leaf size and shape. It is not as strong as the elements of color and form. The latter two are things that can usually be read from a distance while the perception of texture is something that will diminish and disappear entirely as distance increases. Still, texture is somewhat like the "icing on the cake"; it is important to the overall composition and can add a certain richness to the product.

In the small garden or small space where things tend to be appreciated close up, textural contrasts (coarse against fine) and textural gradations (coarse, medium, fine) are quite practical. In those situations where some special plant or structural element is being displayed, a background with a single definite texture can add a great deal to the impact of the display. In small spaces, the illusion of depth, from a single viewpoint (for example, a window), can be also created with texture if a gradation from coarse to fine is used.

Figure 8. The Dynamics of Autumn in a Mixed Landscape

Chapter 2
Planting and Maintenance

Planting Procedures

While the prescribed seasons for successful tree and shrub planting have changed in the last 20 years due to the fact that most nursery plants are now containerized, the horticultural practice has not. Successful planting of woody plants still involves paying attention to the size of hole into which a plant is being placed and the soil in which it is being planted. Coupled with these are a number of common-sense practices involving how to look after the plants before planting, how deep to plant, how to react when soil conditions for planting are either too wet or too dry and how to select a good day for the job.

Nursery Stock

Prior to the advent of container growing by nurserymen, trees and shrubs were purchased either **bare-root** or **balled and burlapped**. Both kinds are still available from nurseries, but by far the greatest number of plants sold today are containerized. Container-grown plants are generally excellent if they have been well cared for prior to sale. All containerized plants, unfortunately, have not been container grown. It is not uncommon for buyers to find that the product they thought was container grown was nothing more than a field-grown plant that had been stuffed into a container shortly before it appeared in the sales yard.

Planting Bare-Root Stock

The nature of the stock purchased will call for some variation in the approach to planting. Fall planting is not generally recommended for the prairie region when bare-root stock is being used. Spring planting is much more satisfactory for this type of material because the plants are assured a longer period of favorable growing conditions for re-establishment. Excavations for plant root systems in all cases must be larger than that normally required for the root mass. When dealing with bare-root material, the roots must be allowed to assume their normal positions and the planting depth should not exceed that at which the plants were grown in the nursery.

Planting Balled and Burlapped Stock

Nursery stock that has been grown for sale as balled and burlapped material will have a very compact root system; consequently, some of the special procedures that are required for the planting of bare-root and container stock do not apply.

The burlap used for wrapping the root ball is usually of a special type. It is coarser than the sort of burlap most often seen, so plant roots are not confined to any significant degree. It is also a material that readily breaks down. Because of this, there is no need to remove or slash the burlap at planting time. It is recommended that the burlap be left in place, but opened at the top to avoid possible girdling where it is tied. All parts of the burlap

Figure 9.
Planting Bare-Root Stock

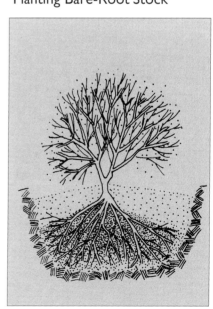

Figure 10. Preparing a Container-Grown Plant for Planting

Planting Containerized Stock

With container-grown plants, the purchaser has far more leeway. Provided such plants are vigorous and air temperatures are not too high, successful planting can be accomplished at almost any time during the growing season. The size of the excavation must be larger than the root mass so that roots will be encouraged to leave the root ball and become established in the surrounding soil. When the plant is removed from the container, examine the root mass to make sure that the root system is not turning in on itself. If any major roots appear to be growing into the center of the ball rather than towards the outside, then carefully cut them out if they cannot be made to face in the proper direction.

Sometimes, if a plant has spent a very long time in the container, it will have had a tendency to become pot-bound. When this has occurred, the fibrous roots will form a thick, heavy mat at the place where the growing medium meets the edge of the container. The roots of pot-bound plants can be encouraged to grow into the surrounding soil by loosening the mass of compacted roots with a knife blade, or even a plastic label, using a gentle combing action.

When the plant is set in its pit, it is advisable to place it on a small mound of soil. This will set the rootball just above any perched water table that can develop at the bottom of the hole should a gardener become overzealous with the hose. If this is done first, and backfilling is handled carefully, the plant will also be assured of good backfill soil in all parts of the root zone.

Planting Depth

There are a few useful comments to be made with respect to planting depth. Plants growing on their own roots should be planted at the same depth at which they were grown in field or container. Plants that have been grafted on a foreign rootstock should be planted so that the graft union is located about 5 cm below the soil surface. This discourages possible top growth from the rootstock. Also it offers some protection to the union itself. Some gardeners, particularly those who grow tender grafted roses, commonly use this technique. However, they have also been concerned with getting the root system too deep. To avoid this, they will lay the rootstock almost horizontally at 5 cm below the surface and then bend the upper part of the plant vertically. By doing this they protect the graft union, discourage the rootstock from producing top growth, and still avoid placing the root system too deep. Fortunately, most of the larger shrubs that have been grafted on rootstocks will tolerate the deeper planting, so this sort of manipulation is not necessary.

When setting a tree or shrub in its new location, place the central portion of the plant 5 cm higher than the edge of the back-filled hole. Applied moisture will then be encouraged to move to where it is most needed and not to the center of the plant where it can, in some cases, contribute to crown rot.

Soils and Nutrition

Do woody plants have nutritional problems? Of course they do, but these are not likely as common as what most people are led to believe. Woody plants are much more likely to have problems resulting from the physical rather than the chemical nature of the soil.

Soil Texture and Structure

There are two properties of soils that all gardeners should appreciate. These are **soil texture** and **soil structure.** Soil texture refers to the percentage of sand, silt, and clay in a soil. This will determine whether the soil is heavy or light. The higher the clay fraction the heavier the soil. Soils for woody plants should not be too heavy. If they are, root systems will suffer, and as a consequence, the above-ground parts of the plant will be adversely affected. Neither should soils be so light that they fail to retain and supply moisture and nutrients to the plant.

Experienced gardeners will always examine the texture of their soil to determine whether or not it should be improved. Should physical improvement be called for, the usual way of going about it is by adjusting the soil structure.

Soil structure differs from soil texture in that it relates to the way the soil particles are arranged. In a topsoil with a good soil structure, the individual soil particles are bound together in clusters or aggregates of various sizes. Naturally-formed aggregates are fairly water stable. A good soil structure provides good drainage and aeration and does a great deal to assure the development of vigorous plant root systems.

While clays make good binding agents, soil organic matter which ultimately breaks down into gums and mucelages influences soil aggregate formation in surface soils even more effectively. This is the reason why sphagnum peat is so highly valued as a soil amendment.

Plant Nutrition

Like all other plants, trees, shrubs and groundcovers require a balanced nutrition for normal growth and development. Nutrient deficiencies, fortunately, are not common but they can and do occur. Because soil-testing services are generally available for a modest fee, it is no longer necessary to guess about possible nutritional deficiencies.

When soil test results show a need for any of the three major nutrients, make fertilizer applications early in the growing season. Woody plants are much slower than herbaceous plants to respond to applications of fertilizer. Nevertheless, if those containing nitrogen are applied late in the growing season, they may very well promote late growth that will not mature at the normal time and therefore not be able to harden up for winter.

Should a woody plant demonstrate a need for some trace element, very often

Figure 11. Planting Grafted Material

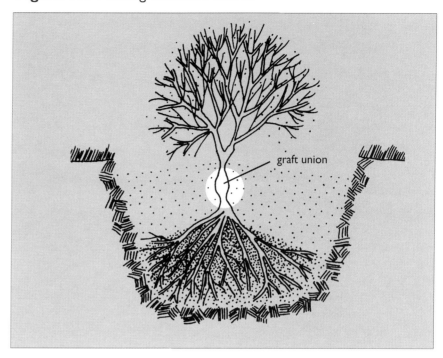

these can be applied to the foliage in spray form, but since the safe limits for application of trace elements to foliage is so narrow, they are best applied with caution. It is also important to know that the major element, nitrogen can also be effectively applied to the foliage should a quick response be needed at the beginning of the season. In such cases, however, the nitrogen carrier must be one that will not burn the foliage.

Soil Acidity and Alkalinity (pH)

In the prairie provinces, the more common nutritional problems with woody plants seem to be those relating to soil reaction (acidity or alkalinity) and salinity. In the drier areas, alkalinity problems involving excess free lime in the soil tend to be common, as is the symptom, lime-induced chlorosis (see Figure 12), which frequently shows up on the younger leaves of plants. When chlorosis occurs, the veins of the leaf will retain their dark green color but in severe cases the interveinal portions will not show any sign of chlorophyll.

While even high-lime soils contain adequate levels of iron, the high pH, caused by the excess lime in the soil, ties up the iron in a form unavailable to the plant. Products containing chelated iron can be applied directly to plant foliage to correct the symptom but will not have any effect on the cause. To correct the basic condition, the soil reaction must be changed. Elemental sulfur is commonly used to lower the pH of the soil and so change iron to an available form for the plant.

In the more northerly, wetter areas, soil reaction (pH) problems are more apt to be those caused by acid conditions. These are not as easy to recognize since they are expressed simply by poor growth and consequently may be associated with any number of things.

Soil Salinity

Salinity problems are most common in the southern parts of the provinces where they are most frequently associated with poorly drained sites. The problems are caused by high concentrations of salts in the soil and their correction usually involves the installation of subsurface drains to remove salt-saturated soil moisture. In the presence of high soil salinity, plant roots are severely stressed. This has an indirect effect on growth of the tops.

Figure 12. Chlorosis

In parts of the region where these conditions exist, it is not uncommon to see accumulations of salts right on the soil surface of low, wet areas in farm fields when drying winds have evaporated the moisture in which the salts were previously dissolved.

Pruning Practices

Pruning at Planting Time

When bare-root material is planted, light pruning is also recommended. The amount necessary will depend on the condition of the plant. If the plant is dormant, the amount called for will be minimal. On the other hand, if the plant is showing active growth, heavier pruning may be necessary. This is because the root system that formerly supported the top has been reduced in digging and can no longer get moisture to the newly-opened leaves fast enough. By reducing the top of the plant, a better balance of roots to shoots will be achieved and incipient wilting avoided. While the objective to re-establish a balance between top and root is still a valid one, it has been shown that attempts to physically re-establish this balance by moderate to heavy pruning is not necessarily practical. Active shoots produce metabolites and growth regulators which are necessary for the development of the root system; hence heavy pruning, while seeming to be a move in the right direction, may actually have the opposite effect.

Pruning at planting time is advocated to aid in the establishment of the new plant. It should be done with the future form and growth of the plant in mind. In other words, when wood is to be removed, only those branches and branchlets that are not or will not be contributing to the form or future well-being of the tree or shrub should be selected for removal.

Trees

With trees, be prepared to retain the scaffold branches. These are major branches that are well distributed throughout the top. In later life they will provide a structurally sound framework for the tree. Smaller, weaker branches can be dispensed with entirely. Other things to be considered for removal are branches growing towards the interior of the tree. Such growth may ultimately interfere with scaffold branches and cause damage to the latter when the surfaces of the two are in direct contact. Finally, some shortening or heading-back of the retained branches may be desirable. If a branch in one part of a tree has grown at the expense of some of the others and is obviously spoiling the shape of the tree, heading-back the offending branch may be the answer.

Shrubs

The pruning of bare-root deciduous shrubs at planting time calls for the same general considerations and procedures used for trees. For most newly-planted shrubs, all that may be required is some light heading-back, coupled with removal of some of the interior wood. This is not to say that the person doing the pruning should be any less selective in choosing what and what not to remove from shrubs. The only difference lies in the fact that pruning mistakes on shrubs are generally far easier to correct than they may be on trees.

Pruning Cuts

There is one other thing about pruning that is very important and that is how and where to make the cuts. The technique is a straightforward one and yet few people, "professional" or otherwise, take the time to learn to do the job properly. Most cut away happily, paying little, if any, attention to where the cuts are made as long as the plant is beginning to conform to the shape they wish to impose upon it.

Cuts are very important to the future well-being of the plant. If they are done well, the natural form of the plant is retained and the cuts heal quickly to

Figure 13. Pruning Practices: Shrubs at Planting Time

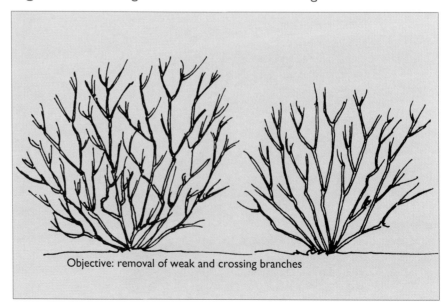

Figure 14. Pruning Practices: Trees at Planting Time

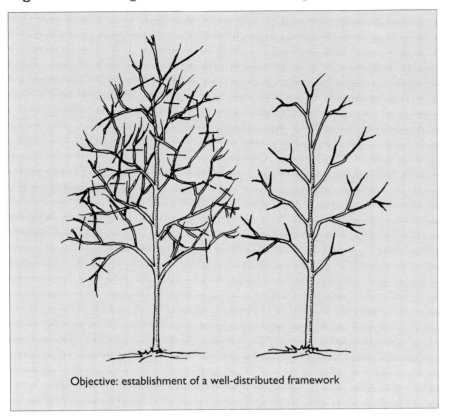

Objective: establishment of a well-distributed framework

Figure 15. Pruning Practices: Types of Cuts

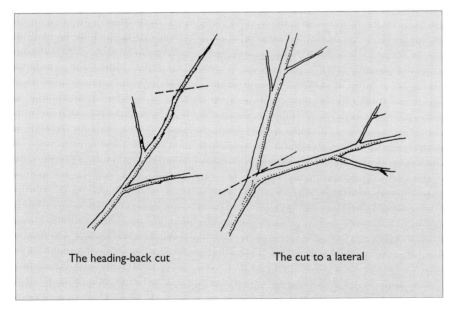

The heading-back cut   The cut to a lateral

prevent any infection by disease or wood-rotting organisms. There are three types of cut. The **heading-back cut** is used to shorten a branch; the **cut to a lateral** is used to redirect growth as well as shorten a limb; and the **cut to a branch collar** is used when complete limb removal is called for.

The heading-back cut is one that is frequently abused. Unfortunately, it is a very common sight to see freshly pruned trees that resemble an old fashioned coat rack more than they do live trees. The purpose of the heading-back cut is to shorten a limb not simply truncate it. For this reason, the cut should not be carried into wood that is more than two years old. With the heading-back cut, cut back to slightly less than 0.5 cm above a bud. When this is done, closure will be effected more rapidly. When heading-back cuts are made without regard for the position of the closest bud, the cuts seldom "heal" before wood-rotting fungi or disease enter the plant via the open wound.

The cut to a lateral is another instance where the cut is too frequently made on wood that is far older than it should be. When such a cut is being made, the lateral should be similar in diameter to that portion of the branch that has been selected for removal and yet it is not uncommon to see branches of 7 to 10 cm in diameter cut to laterals with diameters that do not exceed 2 cm. Avoid such mistakes. When pruning or cutting to a lateral, make the cut about 0.5 cm above the lateral. The alignment of the cut should be approximately parallel to the direction in which the lateral is facing (see Figure 15).

When limbs are to be removed from a tree, make the cut just outside the branch collar (see Figure 16). It has been demonstrated that collar tissues are capable of giving quick protection to open wounds of this magnitude by creating an impenetrable barrier to prevent invasion of organisms even before the generation of callus occurs.

No matter what type of cut is being considered, make one that is going to close properly. This means that each cut must be made at a point where this can effectively take place. It has been pointed out (Shigo, 1982) that the covering of wounds with a tree dressing is neither necessary nor desirable if the cut has been made at the proper place. It has also been noted that most wood-rotting fungi tend to sporulate in the fall; hence fall pruning may not be such a good idea with some plants. One practice that appears to hasten the callusing and natural covering of wounds is the use of a sharp knife to lightly trim and bevel the edges of the wound. For this purpose, use a stout bill-hook pruning knife.

Figure 16. Pruning Practices: Removing Large Limbs

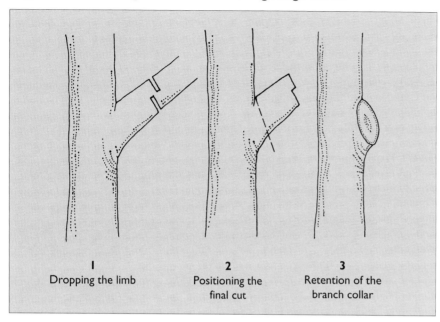

1 Dropping the limb
2 Positioning the final cut
3 Retention of the branch collar

Annual Pruning

Every woody plant requires some pruning on an annual basis whether it be to retain the natural shape of the plant, to remove dead or damaged limbs or to prevent two or more branches from rubbing on one another. If the plant is a flowering one, there is a right and a wrong time to prune. The critical pruning season relates to the time when flower buds are developed by the plant.

Spring-Blossoming Plants

If the flower buds are formed a year prior to blooming, such as they are with lilacs, flowering crab apples, and all other spring-blossoming plants, the right time to do the pruning is after flowering has finished and before the buds for next year's flowers have been formed. If pruning is done at this time, none of next year's flowers will be lost in the process. On the other hand, if pruning is done after the flower buds have formed but before blossoming has occurred, the flowering surface will likely be reduced.

Summer-Blossoming Plants

Not all of the flowering woody plants produce their flower buds on wood of the previous season. There are some, like the snow hills hydrangea, that produce flower buds on wood of the current season. With these plants, the more new wood, the more highly-productive will be the flowering surface. So any management practice (like heavy pruning at the beginning of the growing season) that will foster the production of new growth is going to be the one to use. With the snow hills hydrangea, the common practice to assure maximum flower production is to prune quite heavily sometime prior to the commencement of new growth. Because this plant has really nothing very much to commend it as far as winter appearance goes, most people prune it in the autumn. If the plant had anything of value to offer during the winter, they would undoubtedly wait until spring.

The *spirea* cultivars belonging to the species *bumalda* and *japonica* fall in the same class as *hydrangea* when it comes to annual pruning, but cuts should be much more selective. These plants also have a tendency to provide continuous bloom provided the old flower-heads are removed once they have started to fade. The pruning practice that is recommended, therefore, is one that not simply dead-heads but one that removes a little of the older non-flowering wood in the process. This encourages the production of new flowering wood, and, when this technique is followed, these plants will not only retain their flowering capability but also their neat attractive habit of growth.

Summary

If planting is done well and the pruning strategy is based on sound objectives, gardeners can expect a healthy, hardy tree or shrub to perform well and survive in the general climate of the prairie provinces. To achieve success, the following two general principles must be kept in mind:

- Whatever affects the root system of a plant will in turn affect the top
- Anything that affects the top of the plant will have a corresponding effect on its root system.

Thus, while good planting procedures will operate directly to improve the root system of a plant, pruning practices can either enhance the effect of good planting or work against it.

Figure 17. Pruning Practices: Cutting to a Bud

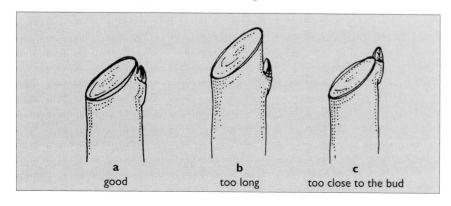

a good
b too long
c too close to the bud

Figure 18. Dieback from Poor Pruning

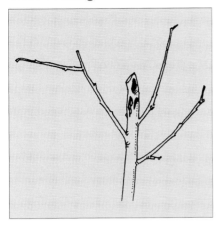

Good pruning is good gardening and it must be practiced with two main objectives in mind:
- To maintain the natural growth habit and health of the plant by removing only that which is necessary
- To promote those qualities for which each plant was originally selected.

Thus, each plant may very well require annual attention to maintain the characteristic or characteristics for which it was selected. If a plant was chosen for its flowering habit, annual pruning must be appropriately timed in order to improve rather than diminish the flowering surface.

Renewal Pruning

Some plants, in particular those that have colorful bark, may require occasional pruning to replenish their brightly colored stems. The Siberian coral dogwood is one such plant and, when some of the older stems have lost their attractiveness, they can be replaced. All that is required is removal of the offending stems right at ground level in the spring. The plant will soon respond with bright colored, replacement growth.

Pruning of Large-Flowered Clematis

Some cultivars of the large-flowered clematis will not perform at full potential unless the appropriate pruning procedure is followed. There are two types of large-flowered clematis: those like the popular C. × 'jackmannii' that produce flowers on new wood only and come into bloom in early midsummer and those like C. × 'Rosy O'Grady' that produce blossoms on both old and new wood, and bloom earlier and for a much longer season.

The two types are managed quite differently. Those that flower only on new wood can be simply cut back hard at the beginning or end of the season. These will produce flowers beginning in early mid-summer. Those that flower on both new and old wood do so over a much longer season if they are managed properly. The management technique used is known as *relay pruning* and involves the following:
a. "Dead-heading" the early-flowering stems to take advantage of their potential to produce the flower buds that will become the earliest blossoms next year
b. Cutting back all late-flowering stems to the ground in late August or early September when they are through blossoming.

As with spring-flowering shrubs and trees, next spring's flower buds for this type of *Clematis* will be produced right after flowering has finished; hence, gardeners should have no difficulty in identifying the stems that are to be simply "dead-headed" and not cut back. Because flower buds on these stems must survive the winter, you are well advised to lower them from the trellis and lay them on the ground to take advantage of the insulating value of snow cover.

Pest Management

Gardeners, it seems, are constantly waging war with one pest or another as they strive to maintain the health and appearance of their plants. Unfortunately, gardeners are not always too well informed as to the most appropriate methods of control, and many of the problems which they sought to eliminate are often aggravated by the approach taken.

Since the 1940s most gardeners have placed a heavy reliance on the use of chemicals to control pests. This has resulted in a reduction in the use of many other practical control procedures and two very serious consequences have emerged:
- Pests have developed a resistance to the chemicals
- A degree of environmental contamination has occurred as a result of overuse of the pesticides.

The effect of this reliance on chemicals has been very serious. For example, by 1975 the majority of the more serious agricultural insect pests in the state of California had developed resistance to one or more major insecticides (Harris). This build-up in pest resistance caused growers to increase both the concentration of insecticide and the frequency of application. The result was environmental contamination.

Other problems have emerged from primary reliance on the use of chemicals for pest control. When broad-spectrum insecticides are used, for example, they frequently cause an immediate marked decrease in a pest population only to have it dramatically rise later to even higher levels. This happens when the pesticide not only kills the target population but also eliminates its predators. When only chemical control measures are used against spider mites, this is a common occurrence. Even when predators are not killed out entirely by the pesticide, many of them starve or emigrate from the immediate area in search of food. This permits the surviving spider mite population to multiply rapidly since its food supply, unlike that of the predators, has not been affected.

Because no single control measure has yet been found to be totally effective on pests, and because there has been over the years a greater understanding of the dependence of living organisms on one another, the approach to pest management has changed. Recognition of the fact that control over the population of one species will have an effect on other components of the

Figure 19. Effect of Leaf Miner on Birch

system has led to the the adoption of a concept which accepts certain levels of the pest population. Provided these do not exceed an economical or aesthetically acceptable threshold, no action is called for, but when a pest population reaches what has been called the "control threshold," concerted action is taken. This approach to pest control is referred to as integrated pest management.

For landscape plantings, integrated pest management has adopted the following premises.
- A plant or a landscape planting is a component of a functioning ecosystem; therefore, actions must be designed to develop, restore, preserve, or augment natural checks and balances, but not necessarily eliminate the pest species
- The mere presence of a pest does not constitute a pest problem. Acceptable population and damage levels must be determined
- All possible pest control options should be considered before action is taken. Techniques employed should be as compatible as possible.

Most home gardeners have a control threshold that is arrived at when the aesthetic threshold is reached. Such a threshold will be exceeded more readily by a plant that is normally viewed close-up than one that is viewed from a distance, even though the level of infestation may be the same in both cases. When the aesthetic threshold has been reached in a small private garden, action to control the pest is usually justified.

In larger landscape constituencies like cities and provinces, control thresholds are more stringent and are usually established to counter the more serious problems. In such cases, the plants that are affected must exist in numbers that will make the effort cost-effective and, secondly, the plants must be particularly valuable. A good example of such a control threshold would be that set up to counter the spread of Dutch elm disease.

With integrated pest management, there is a wide variety of control options that can be used in addition to the chemical one. Many of these are not new but they do have special value if they are used in an appropriate combination or sequence. Management strategies that are preventive, for instance, like practicing good sanitation, maintaining plant vigor, pruning, and cultivating, all have a place in integrated pest management. There are also a number of more sophisticated options. The genetic option is one that has been used for some time in the development of plants that have resistance to both disease and insects. Also, there are programs that have been developed to eradicate a population of a destructive insect species by the introduction of sterile males. There have also been some particularly innovative cultural options designed to attract insect predators.

Currently we have trees with problems that would appear to lend themselves to the more comprehensive approach of integrated pest management. In many parts of the region, for example, birch trees have been infested with birch leaf miner (*Fenusa, Profenusa,* and *Hetararthrus*) and within cities at least, the damage has been so severe that the aesthetic threshold of home owners has already been exceeded. Unfortunately, a totally satisfactory solution short of cutting down the trees has not been found.

The ultimate remedy for this problem is one that is not going to be found easily. A lot of options must be explored. For instance, why is the infestation different in residential areas than it is in forested areas? Do we have any plant selections that appear to be resistant to the pest? Does the pest have predators? If so, what are their environmental preferences? And so it goes; integrated pest management is a comprehensive approach to problem solving and one that might well be used with other environmental problems.

References

Harris, R.W. 1983. *Arboriculture.* Prentice-Hall: Englewood Cliffs, New Jersey.

Shigo, A. 1982. "Tree Decay in Our Urban Forests: What Can Be Done About It?" *Plant Disease* **66** (9).

Chapter 3
Woody Plant Problems

The Prairie Region

For the gardener, success with trees and shrubs is very much dependent on a knowledge of regional conditions. No single horticultural region in Canada is without its problems; hence it is often necessary to devise appropriate strategies to foster optimal plant performance. While the need for such measures may seem unimportant in certain parts of the country, the prairie region is not one of them. The growing of woody plants in most parts of the prairie provinces continues to be a challenge.

Inter-Regional Climates

If the country as a whole is thought of as a land mass containing a number of climatic regions, the three prairie provinces stand together as far as climate is concerned. The general climate is one of short growing seasons with moderate to low rainfall, and severe winter temperatures with moderate to light snow cover. Because of these factors, many plants that are indigenous to such places as coastal British Columbia or southern Ontario cannot be expected to perform well or even survive on the prairies.

While most gardeners have seen plants injured by winter conditions, not all realize that it is the length of the growing season on the prairies that is the main factor limiting the variety of woody plants that can be successfully grown. Hardy plants will react positively to the subtle but important changes in day-length of the late growing season by first maturing their tissues and then cold-hardening them through exposure to the steadily decreasing temperatures of autumn. Only when these two conditions have been met can woody plants be expected to withstand the very low temperatures of a prairie winter. On the other hand, introduced plants from places where the growing season is much longer can't react to the region's seasonal cues early enough to mature their tissues and therefore are unable to prepare for winter to the same degree. This is the one of the main reasons why woody plants from other parts of the country cannot always be expected to survive in the climate of the prairie provinces. Nevertheless, gardeners should still try plants from places where the climate and length of growing season are somewhat different. Even though the chances of success may be remote, occasionally a species or cultivar is capable of making it in the new and more demanding environment. When that occurs, the sense of achievement can be great.

Intra-Regional Climates

While there is a general climate for the prairie region, the grower of woody plants will be quick to recognize some quite marked differences within its boundaries. Generally speaking, in the settled parts of the provinces there is a marked climatic difference existing from north to south. The wetter, cooler climate is to be found in the northern areas with warmer drier conditions existing to the south. This is also reflected in the vegetation pattern. The boreal forest existing on the northern edge of major settlement gives way to the aspen parkland. It, in turn, merges with the dry to droughty environment of the grasslands which carry on southward to the international boundary and beyond.

The climate of the southern part of the region is characterized by the presence of drying winds which permit little effective snow cover during the winter. The many hours of sunshine and warm, if not hot, summers also contribute to the dryness. Because of these factors, the gardener in the south must be an even more superb strategist than his or her northern counterpart in order to achieve success with woody plants. In the south, microclimates that provide year-round wind protection and snow cover are an absolute necessity with all but the hardiest of species.

Zonation maps for woody plants, like those published by Agriculture Canada, are a reflection of the regional climate. They show the variability that exists in the different parts, based on the length of the frost-free period, the mean minimum temperature of the coldest month, the depth of snow cover, and the effect of wind. Zonation maps, therefore, are of great value in illustrating the variability that exists in regions as large as this one.

While it is important to recognize that woody plant survival is dependent on the nature of the climate and/or on the use or creation of an appropriate microclimate, there is one additional dimension of climate that can be important to the gardener. It is the dimension that lies between the other two and is referred to as the **mesoclimate.** The mesoclimate is defined as the climate of a relatively small area but larger than that involved when speaking of microclimate. A mesoclimate might be that influenced by the proximity of a small lake kept open in the winter by the hot water discharge of a large power plant. Likewise, the climate within the confines of a large city can fall into the category of mesoclimate. Plants will generally benefit from a favorable mesoclimate. The heat of the city will frequently delay the onset of freezing temperatures at a time of year when the woody plant is still in the process of maturing its tissues. Plants of borderline hardiness which require a longer period to mature will be favored by this aspect of mesoclimate. Less winter injury can be expected because maturation is more likely to proceed to completion under these conditions.

The heat of the city can also be of benefit to woody plants in another way since it can reduce the possibility of extreme low temperature that may cause injury to cold hardened tissues.

Problem Types

Low Temperature Problems

Of all the environmental factors to which plants of the region are exposed, low temperature remains the least controllable. As might be expected, the effects of low temperature will vary with such factors as plant species, stage of growth, age, condition, and type of tissue.

To make the appropriate adjustments for winter survival, hardy woody plants cease active growth sometime before the end of the growing season. When this occurs, the metabolic activity involved with growth, changes, and the plant enters a period of profound inactivity. This condition can be terminated naturally only by exposure to low temperatures (about 7°C) for a specific period of time. Once this period (dormancy/rest) has been completed, a plant can then be expected to resume growth as soon as growing conditions return. Because these seldom occur for several months, the plant remains inactive. This particular stage of inactivity is referred to as dormancy/quiescence since it is imposed only by environmental conditions. Many knowledgeable gardeners will react to this by taking branches of their favorite flowering tree or shrub in early January, after dormancy/rest has been completed, and forcing the flower buds to open. This can be done by bringing the cut stems into the warmth of the house, placing their bases in water, and then waiting a few days for the buds to open.

Because of the severity of the winter climate, many of the early prairie horticulturists sought woody plant materials that were indigenous to places with climates at least as severe as their own. As a consequence many of the plants now commonly grown here have been derived from progenitors that came from such places as Siberia, North Korea, Manchuria, and the colder parts of eastern Europe. Most plants from these places have demonstrated the inherent ability to mature and develop cold hardiness early in the autumn and retain it long enough to withstand the critical low temperatures of winter. In spite of these successes, however, winter hardiness continues to be a never-ending subject for study and research in the prairie provinces.

One of the complications involved with plant hardiness is the degree of susceptibility to low temperature injury shown by adjoining parts of a plant. Roots, for example, are the most susceptible, followed by flower buds, shoots and shoot buds, and finally mature wood. It has also been shown that adjoining tissues may vary in their degree of susceptibility to low temperature.

Plants can be damaged or killed by low temperatures at almost any season of the year. The most critical periods are:
- In mid-winter following the end of dormancy/rest
- In the autumn before cold-hardening has developed to its fullest extent
- In the coldest periods of winter.

The types of low-temperature injury to be expected from time to time in this region include the following:
- Loss of flower buds
- Sunscald
- Frost splitting of tree trunks
- Black heart
- Tip-killing
- Winter-burn of conifer foliage
- Blossom damage from late frosts
- Injury to foliage of broad-leaved evergreens
- Loss of foliage and shoot growth by deciduous trees and shrubs
- Root injury.

Loss of Flower Buds

This type of injury has been common on such things as the double-flowering plum, *Prunus triloba* 'Multiplex'. At one time it was quite common for this plant to lose all flower buds that were carried above the snow line and yet show complete winter hardiness of roots, shoots, and shoot buds. Fortunately this has improved with time. In most urban areas, the flower buds of the double-flowering plum are now seldom damaged, which leads one to suspect that the heat of the city is moderating the late winter temperature. When this type of low temperature injury occurs, it is generally following the end of dormancy/rest.

Sunscald

This problem most frequently occurs on the southwest sides of tree trunks when the winter weather has been bright and sunny. When such conditions occur, ice, which normally can exist safely in the inter-cellular spaces of the bark and cambium without causing injury, melts and begins to re-enter the cells. When this happens late in the day, the west or south west side of the tree experiences a sharp drop in temperature as the sun sets. Water is unable to leave the cells quickly enough. Ice crystals then form within them, rupturing membranes and causing death of the affected tissue. Smooth-barked trees are subject to this type of low-temperature injury.

Fruit growers have learned to cope with sunscald. The common practice is to plant the trees so that the lowest branch comes off the trunk not more than 45 cm from ground level and faces southwest. Such a branch when trained more or less horizontally will do an effective job of shading the trunk from the rays of the late afternoon sun during the winter. Trunks have also been painted with white latex paint to effectively reflect the rays of the sun and moderate daily temperature extremes.

Frost Splitting of Tree Trunks

This phenomenon is the result of the outer wood and bark contracting more rapidly than the inner wood when extremely low temperatures occur. When the tensile strength of the outer tissues is exceeded, longitudinal cracks in the bark and outer wood of some tree species occur. This type of injury can lead to infection and to the ravages of wood-rotting fungi. Generally the tree will produce callus from tissue on both sides of the split and these will grow together; however, when low temperatures return in subsequent years, the split will reopen. The Amur cherry, shown in Figure 20, is frequently injured in this way. Green ash trees, too, almost always have a frost crack.

Black Heart

It has been found that adjacent tissues of plants can show differences in hardiness and that these differences can change, depending on the time of year. In stems, for example, the living cells of the xylem are hardier than those of the phloem in early winter but are several degrees less hardy than the phloem and cambium in mid-winter. Should severe temperature conditions occur at this time, the smaller branches of trees and shrubs may show the symptom of black heart, a blackened ring of damaged xylem tissue. This condition can result in poor growth, reduced flowering, and possible death of shoots.

Tip-Killing

Winter injury to shoot growth on both trees and shrubs is the most common type of low-temperature injury. It may be confined to the tips of new growth or can go as far as the killing of all above-ground parts. When tip-killing is limited to a few inches of the previous season's growth, the usual diagnosis is "for some reason the plant did not cold harden its tissues completely for winter". Such symptoms may occur in consecutive years and may well be cause for concern. In many cases, however, plants will completely

Figure 20. Frost Splitting of Tree Trunks

"outgrow" the condition after a year or two.

When tip-killing exceeds a few centimeters of the previous year's growth, there is serious cause for concern. In such cases, the diagnosis is that the plant is very likely not suitable for the region. Occasionally, some of these plants may have some value; the tree of heaven (*Ailanthus altissima*) has been grown in Edmonton but will kill to the ground every winter. Each year, after it was cut back, however, it would respond by producing an abundance of luxuriant foliage which more than paid for its keep.

Winter Damage to Conifer Foliage

The exposed foliage of conifers, particularly spruce, arborvitae and juniper, is often damaged by winter conditions (see Figure 21a). The damage, a browning or reddening of

Figure 21. Winter Burn of Conifer Foliage

Woody Ornamentals

17

the foliage, is frequently referred to as "winter-burn". It is a symptom of desiccation brought on by the combined effect of very low temperatures and strong winds. These symptoms are very common on conifers in the chinook belt when unseasonal temperatures and high winds occur while the soil is still frozen.

Blossom Damage from Late Spring Frosts

This type of injury has always been a cause for concern by fruit growers because a killing frost during the time of blossoming or fruit-set can make the difference between having and not having a crop.

Injury to Foliage of Broad-Leaved Evergreens

The foliage of such plants as Oregon-grape (*Mahonia aquifolium*) and Oregon-boxwood (*Paxistima myrsinites*) is damaged when exposed to bright sunlight just at the time the snow is beginning to disappear. This is nothing more than desiccation injury caused when transpiration of moisture from the foliage exceeds that being taken up by the roots. Such a condition prevails when the soil is still frozen. When these symptoms occur they are a positive indication that the affected plant or plants are growing in the wrong microclimate and therefore must be moved to a spot where the condition can be avoided.

Loss of Foliage and Shoot Growth of Deciduous Trees and Shrubs

In some years when spring growing conditions are abruptly interrupted by the return of winter, the new growth of trees and shrubs may be totally destroyed. Fortunately, most healthy plants have the ability to replace their foliage. Many trees, for example, are capable of producing new growing points from the base of the damaged tissue so that a second flush of growth can take place.

Root Injury

Root injury is probably the most interesting of all temperature-related injuries to plants because of the nature of the symptoms. When root damage occurs there is no outwardly visual symptom. The plant appears normal until the start of spring growth and even then the swelling of buds and unfolding of leaves proceeds normally, up to a point. Just before the leaves start to size-up, however, the growth process stops, and newly formed leaves wilt and dry up.

The explanation for this is that the opening buds had sufficient nutrients, growth substances, and moisture available to them, from their own and adjoining tissues, to commence growth. Only when the damaged roots were called upon to perform their usual function for continued growth were they unable to do so and the new growth died.

Plant Diseases

Like every other living thing, woody plants are subject to attack by disease organisms. There are four main classes of these: **fungi, bacteria, viruses,** and **mycoplasmas.** Each can cause serious problems or even death of the plant infected. Generally speaking, stressed plants are more susceptible to disease than healthy ones. However, if a plant is a susceptible host, if the organism is aggressive, and if soil and climatic conditions are favorable, even healthy plants may be affected.

Of the four disease types, those caused by fungi are the most common. Plants are more susceptible to this type of parasite since the spores which germinate on the surface of the host do not require any special point of entry but may simply invade through healthy tissue. Bacterial infections, on the other hand, must have direct access to the internal tissue. Entry is common through open wounds but may also be made via flower nectaries and leaf stomata. Viruses are unique in that they reproduce only within the living plant. They are chiefly transmitted from infected to healthy plants by sucking insects, though they have also been transmitted by grafting and pruning practices. Mycoplasmas also multiply in living cells and, while they have limited mobility within the plant, they can move through the phloem. The most common vector for mycoplasma infection is the leaf hopper which only becomes capable of transmitting the disease after the organism has been incubated for a certain length of time within the body of the insect.

Fungal Diseases

Something that has only recently become a problem in domestic plantings is *Armillaria* **root rot.** While this fungus disease complex has been known to attack other woody plants, it has become a major concern in the prairie provinces because of its effect on the very attractive and popular Amur cherry. The main organism responsible is *Armillaria mellea*, the fruiting body of which is the popular honey mushroom.

Amur cherries infected with *Armillaria* root rot show a general loss of vigor and go into a period of early senescence. It is not uncommon to suddenly find 15-year old trees showing a loss of vigor and a marked discoloration of the attractive golden bark. The most reliable signs of the fungus are fan-shaped mycelial plaques to be found between the bark and the wood at the base of the trunk. Also dark root-like fungal structures that resemble shoe strings are to be found on roots just below the soil surface. These are capable of infecting healthy roots on contact. There is no cure for the disease, but it is recommended that trees be kept from encountering stress and, when a diseased tree is removed, it should be replaced with a resistant kind. It has been shown that the progress of the disease can be slowed by exposing the root collar and larger roots for a distance of 60 cm from the trunk to permit the removal of rotting roots and infected bark and soil.

Dutch elm disease, the most destructive disease to affect woody plants in this century, has been identified throughout Manitoba and in native stands of American elm in the Qu 'Appelle Valley of Saskatchewan and other south-eastern parts of that province. Fortunately, the bark beetles which have been vectors for the spread of the fungus in eastern North America are not native to western Canada. However, federal forestry officials report that a native bark beetle occurs in the range of the American elm in Saskatchewan and Manitoba, and is considered a possible vector for transmission of the disease-causing fungus.

Healthy trees are usually infected in spring and early summer when the bark beetles feed on one- and two-year-old wood. The beetles make wounds and, in the process, transfer spores of the fungus to the sapwood. The first observable symptom is the sudden discoloration and flagging of the leaves at the tips of one or more branches. Leaves turn yellow and roll up at the edges and finally turn brown, remaining on the tree. Some trees infected in early summer may be completely killed the first year while others may last for several years.

No elm species has been found to be completely immune to the disease. Although Asiatic species were thought to be fairly resistant, they have not proved to be so.

Bacterial Diseases

Of the woody plant diseases in the prairie provinces, **fire-blight** is the one discussed most frequently. Caused by the bacterium *Erwinia amylovera*, this disease attacks only certain plants of the rose family (*Rosaceae*) including apple, pear, crab apple, mountain-ash, and cotoneaster. It does not attack such things as ash, birch, and pine trees, as many amateur gardeners might have us believe.

Fire-blight is a disease that lives over in blighted twigs and cankers. In the spring these may produce a sticky exudate which is picked up by pollinating insects and transmitted to other trees. The bacteria may gain

Figure 22. Dutch Elm Disease

entrance to the tree via the nectary of the flower. In a short time the clusters of blossoms and leaves at the site of the infection wilt and turn black. Should rainy weather occur at this time there is real danger of the disease being spread from the upper to the lower portions of the tree. The primary infection is commonly referred to as **spur blight** since in most instances it is confined to the infected blossom clusters. Should the disease advance from the sites of primary infection, and it is capable of doing so, then **stem blight,** characterized by spindle-shaped swellings on the branches, will occur. The swollen sections, or **cankers** as they are often called, should be removed before the start of the growing season and burned.

Fire-blight, in a bad year, can result in serious damage or death of an infected tree. Whenever spur blight is noticed, take action immediately. Pruning can be effective in preventing the spread of infection provided cuts are made into healthy tissue, at least 30 cm below the site of infection. Since the disease can be easily transmitted by pruning equipment, it is also very important to disinfect the pruning tools with an effective bactericide after each cut is made. All material removed from the tree should be burned immediately to avoid re-infection.

Sprays of the bactericide *Agrimycin* have been used to protect apple and crab apple trees at the time of blossoming. Also some cultivars of these trees appear to be less

Figure 23. Fire-blight

susceptible to the disease than others. For further information on the use of spray materials contact your provincial Department of Agriculture.

Viral Diseases

Viral diseases of woody plants are rarely fatal although they are quite capable of seriously weakening a plant or causing dwarfness. The more common symptoms of viral infection are distorted growth of plant parts, inhibition of vegetative growth and suppression of chlorophyll. There has been a good deal of work done with the viral diseases of fruit plants, but very little with landscape plants, although many of the viral diseases attacking fruit plants also attack the others. Agriculture Canada has an effective program to prevent the importation of virus infected plants from other countries.

Mycoplasmal Diseases

The most common mycoplasmal disease of woody plants is the one known as **silver-leaf.** It is the name reserved for diseases caused by *Stereum purpureum* and a few close relatives. It is very common on such things as cotoneaster and is easy to recognize by the dull silvery color of the foliage. There is apparently no cure for the disease which generally weakens and discolors the plant. Treatment is similar to that for fire-blight.

Figure 24. Forest Tent Caterpillar

Figure 25. Effect of Scurfy Scale on Cotoneaster

Insect Problems

Woody ornamental plants are subject to a variety of insect and related pests. For easy classification it is customary to lump insects into two groups: those with biting mouthparts which do their damage by chewing, and those with sucking mouthparts that are capable of penetrating plant tissues and sucking the juice.

Chewing Insects

Insect larvae and adults with chewing mouthparts feed on various plant parts. Caterpillars eat holes in the leaves or feed along the margins. Leaf miners feed between the two leaf surfaces and skeletonizers feed between the leaf veins, rasping off the upper epidermis. Adult insects such as beetles, grasshoppers, and leafcutter bees work on above-ground parts, principally leaves.

The most troublesome and the most common chewing insects in the region are the **forest tent caterpillar** (*Malacosoma disstria*) and the **cankerworm** (*Alsophila pometaria*). Both are leaf feeders and attack a wide variety of plants. The forest tent caterpillar is well known for its preference for the leaves of trembling aspen, although it will by no means stay away from other plants when its preferred food source has been exhausted. Cankerworms are serious pests in both Manitoba and Saskatchewan and are quite serious on such trees as American elm and green ash. Fortunately, cankerworms have not yet invaded Alberta. The biologically acceptable spray using the bacterium *Bacillus thuringiensis* is effective on all caterpillars. Although chemical sprays like malathion, diazinon, and methoxychlor are also effective, they can also reduce the population of natural predators. Because of this, insecticidal soaps have also been used. These have been found quite effective in most cases and were less harmful to non-target species. In some parts of the region, home-owners have resorted to measures designed to prevent wingless adult females of the cankerworm from climbing to the top of the tree. The technique uses a band of "tanglefoot" which immobilizes the insect en route to her rendezvous with the winged males.

Three other chewing insects that are sources of concern to people in the prairie provinces are birch leaf miners from the genera *Fenusa*, *Heterarthrus* and *Profenusa*, the pear slug, the larval form form of the pear sawfly *Caliroa cerasi*, and those sawfly larvae that attack some species of spruce and larch.

Only the birch leaf miner is difficult and expensive to control. Because it feeds within the leaf, no inexpensive means of control has been found, even though the systemic insecticide, dimethoate, had been expected to provide the solution. It is important to know that not one, but three, kinds of leaf miners are responsible for the damage so that some measure of control can be achieved. It had become customary to treat the trees early in the season after the leaves had formed and hope for the best. The problem that arose, however, was that if the insecticide was applied early in the season it would probably control the first kind of leaf miner but would not likely be present in high enough concentration by the time a second type appeared. Application of dimethoate as a soil drench has been more effective than bark applications but unfortunately it still cannot be relied upon to give complete control.

In recent years the scale insect, *Chionaspis furfura*, has caused a lot of serious damage to the popular hedge plant *Cotoneaster lucidus*. Unfortunately, in most instances the insect is seldom observed before noticeable damage has been done to the plants. The insect, commonly referred to as **scurfy scale**, feeds in clusters along the stems of the host. The female scales are pear-shaped, greyish and appear to be immobile. Males are white but are much more slender and smaller. When damage is as severe as that shown in Figure 25, then there is no recourse other than to replace the plant or to cut it back to the ground and let it start over. In cases where the damage is noticeably less than this, then one or two applications of a horticultural oil applied during the dormant season should control the pest.

Sucking Insects

Sucking insects attack all parts of plants but prefer leaves and developing shoots. In the process of feeding they are known to stunt and deform new growth, curl leaves, form galls and seriously weaken plants. In some cases, they will actually cause discoloration of the foliage. Sucking insects, frequently, are also the vectors for many serious plant diseases.

Of the sucking insects that are serious pests on woody plants, **aphids** and

Figure 26. Damage to Ornamentals Caused by Aphids

mites are the two most common and most likely to create problems. Aphids attack a great variety of plants. When predators are not present, an application of a broad-spectrum insecticide like malathion gives effective control. With some deciduous plants, however, complete removal of an affected branch could be a simpler way of handling the problem, particularly when there is a danger of destroying natural checks and balances.

The insect damage shown in Figure 26a consists of the galls, frequently seen on the male trees of green ash. These are caused by the ash flower gall mite *Acerica fraxiniflora*. Figure 26b shows the severely distorted growth on honeysuckle caused by the honeysuckle aphid *(Hyadaphis tataricae)*, and Figure 26c shows an aphid population on a stem. Often when these situations are encountered, it is easier to remove the stem rather than spray the plant with insecticide.

References

Blenis, P., Hiratsuka, Y., and Mallet, K. 1987. "Armillaria root rot in Alberta," *Agriculture and Forestry Bulletin*, University of Alberta **10(1)**: 4–5.

Davidson, J.G.N. 1987. "The Principal Diseases of Commercial Saskatoons," *Agriculture and Forestry Bulletin*, University of Alberta, **10(1)**: 6–8.

Ives, W.G.H and Wong, H.R. 1988. "Tree and Shrub Insects of the Prairie Provinces." *Information Report NOR-X-292* Canadian Forestry Service.

Chapter 4
Planting Design

Planting design is not garden design. The two, while involved with the same general subject area, must be approached quite differently. Gardens are designed primarily for people and their activities, while planting design deals exclusively with the arranging of plants in both an ecologically and aesthetically acceptable fashion as elements of the garden. Good planting design enhances good garden design but will do nothing to improve poor garden design.

There are several facets to planting design. The first is planning, since it involves investigation and evaluation of the environments existing in the various parts of the site. The factors that are important here are chiefly soils and light conditions and the things affecting them. Next comes plant selection, that is, choosing materials that thrive within these environments and that will at the same time meet the physical requirements set out in the conceptual plan of the garden. The third and final phase is one that might well be described as "fine tuning", where both plants and their arrangements are examined from the point of how well they meet the purely aesthetic aspects of design.

Planning and Design

Investigation

It may be quite revealing to those who are gardening for the first time to discover just how many microclimates there are to be found even on a small property. For example, areas on the north sides of buildings always create problems because of shade. Sun loving plants and most turfgrasses never do well in these environments and usually give way to weedy volunteers. Areas on the south side of a building have exactly the opposite type of exposure; hence, it is important to think not only of plants that thrive under these conditions during the growing season but also about those that will not suffer from an unusually early start in the spring. For example, there are several broad-leaved evergreens that would not favor south-facing environments because very early exposure to bright spring sunlight would result in severe leaf burn and loss of leaves. There are also locations that may be strongly affected by wind. Such places are always a problem, because if the winds of summer don't make it hard for certain things to survive, the winds of winter almost certainly will.

Microclimate, then, plays an important role not only for the growth and well-being of hardy plants but also for survival itself. This is a factor that is often overlooked and some gardeners are very quick to attribute failure to a lack of plant hardiness. Such failures should always be questioned, but it is of fundamental importance to realize that even the hardiest plants have environmental preferences.

Plant Selection

Once the investigative aspects of planting design have been dealt with, a good starting point might be the choice and placement of trees. During the conceptual phase of garden design, certain things will have been considered, relative to the comfort of people. For example, because people may choose to use a deck more frequently at one time of day than at another, some thoughts relative to the choice and placement of trees will have been already made. If sunlight is preferred during the time of most frequent use, the designer will have made sure that adjoining shade trees were not located in inappropriate spots.

When choosing a tree there are many questions to be considered. How tall will it be? Does it carry its lowest branches at eye level or below (low-headed), or does it carry them well

Figure 27. Tree Form

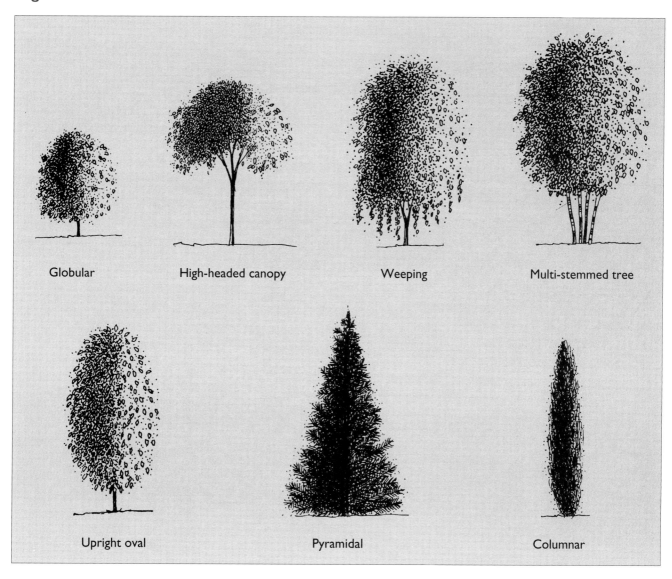

above the head of the average person (high-headed)? What sort of rain shadow will it cast? What are its undesirable characteristics? Can these be tolerated?

Trees are also used in the garden for many purposes other than environmental modification (see Chapter 1). They can be used to break up the mass of sky overhead, and thus create a more intimate environment on the ground. Trees can be used to create that interesting and variable silhouette called skyline, and they can do a very effective job of enframing those things that are to be seen in the distance. They will also do an effective job of screening some visual intrusion from an adjoining property.

Trees have many ornamental attributes that may affect choice. The color of deciduous trees, for example, comes in a wide selection of greens which can be used to advantage. Small differences in the shade of green can have surprisingly significant landscape effects during the growing season. Some trees put on an outstanding display of foliage color in the fall, while others, usually horticultural selections, may display a color other than green the whole season long. The latter should be used with restraint in situations where the natural aspects of the landscape are highly valued; however, when carefully used they can provide points of interest or emphasis in the landscape.

Tree form is also something to be considered. There are quite a number of forms available and, as might be expected, they are not always interchangeable. For example, it would not be possible to replace a high-headed shade tree with a narrow columnar type and not expect a major impact on the landscape. Tree form too can be quite evocative; for instance, the tall narrow columnar form tends to attract the eye and direct vision upwards. Associate this form with a horizontal mass and the impact of the vertical is accentuated still more. The more common globular form, on the other hand, is very static simply because neither its horizontal nor its vertical axis dominates. Because of this

Figure 28. The Weeping Willow and Water: Complete Harmony

feature, globular forms are used to create stability. In other words, they are good anchors. This is particularly true of globular shrubs.

One of the more unusual tree forms is the one with pendulous branches. The weeping willow and the weeping birch are good examples. The weeping willow goes beautifully with a smooth, tranquil pool of water particularly if some of the branches are dipping into it. This is likely because the billowing branch and foliage habit of the tree symbolizes the fountain-head, the source of the water. The weeping birch does not seem to go quite as well with water as the willow because birches are not normally associated directly with water and wet places; nevertheless, the weeping birch is an elegant tree when seen in association with an extensive greensward.

Since gardens and landscapes are to be enjoyed the year round, conifers have a great deal to offer, particularly to the winter scene. However, one problem with conifers is that some of them carry their branches right to the ground (spruce and fir, for example) and have that uncontrollable habit of increasing the basal plant circumference as they get older. Because of this characteristic, spruce and fir must be carefully sited so that the basal growth of the branches does not ultimately encroach on buildings, sidewalks, or other plants. Also, there can be no justification for the destruction of the natural form of such trees which will occur if the lower branches are removed. Fortunately, not all conifers are like this; most pine trees, for example, naturally self-prune as they get older, so removal of lower branches will not destroy the ultimate form of the tree.

One other important thing about the use of conifers is that they should never be mixed with deciduous trees (see Figure 29a). If this is done, the faster-growing deciduous trees invariably grow into the tops of the conifers causing distortion of the latter. This is not to say that conifers and deciduous trees are totally incompatible; they can be used together quite successfully as long as mixing is avoided. Figure 29b illustrates a better arrangement of the two in landscape situations.

Because of their form, conifers tend to be accent types. If they are not used too freely, spruce, pine, and fir can punctuate the landscape, directing the eye to a point and stopping it, or they can create a rhythmic pattern by virtue of their dominant form, their placement, and repetition within the landscape.

Flowering trees are also very popular in the garden landscape. Most of those used are spring flowering types with blossoms in tints of white, pink and red. None of them are massive, and hence are well suited to both large and small properties.

The selection and placement of trees can be an important first design step. However, the designer must always be conscious of the consequences of overuse. Too much shade may affect the appearance and vigor of the lawn. Rain shadows created by tree canopies may make it difficult to grow certain plants within their influence. Tree roots may limit the space available to other plants. Flowering trees usually produce fruit which sometimes has consequences relative to maintenance operations. The Ussurian pear is one such tree. It is a beautiful flowering specimen, but starts to drop its fruit in mid-August. This creates problems, particularly when the canopy overhangs a regularly mown lawn (Figure 30).

Figure 29. Combining Coniferous and Deciduous Trees

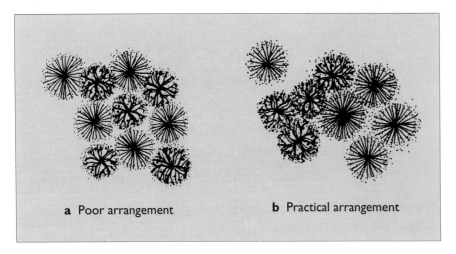

a Poor arrangement   b Practical arrangement

Figure 30. Maintenance Problems

Figure 31. Size and Scale of Plants for the Small Property

Size, Form, Color, and Texture

One of the more important concerns that must be addressed in the early stages of planting design is the selection of shrubs based on their size, form, color, and texture. **Size** and **form** are particularly important in a small garden. No one wants to grow plants that are going to either interfere with other things or get so large they soon have to be removed. For this reason, 75 to 80 percent of the shrubs used in a city garden should never exceed 70 cm in height. Of the 20 to 25 percent remaining, all should be narrow or at least somewhat leggy, otherwise they will occupy more space than what can easily be accommodated. Leggy plants are recommended because smaller things including groundcover can generally be grown beneath them.

On larger suburban or rural properties a designer has much more freedom in the selection of materials. Selection on the basis of size and form is not nearly so restrictive, except perhaps in those smaller more intensively developed areas adjacent to the house.

The **color** of materials is as important to planting design as it is to any other design product. Working with color can be particularly challenging since in so many cases the foliage color of plants changes with the season. Plant color in the spring, for instance, may be quite different from that which is displayed during any of the other three seasons. The changing light at different times of day, rain, snow, fog, frost, and even wind will affect color perception.

Not all color characteristics of plants are of the ephemeral type. Some foliage colors are reasonably constant and unchanging. The silvery greens, which are so frequently found growing in the hotter, drier parts of the region, change very little, if at all, with the seasons. Also, many horticultural cultivars are offered for sale because of some unusual foliage color. In most cases colors of these are also static. As a result, the plants are used more frequently in places where a strong permanent accent is required.

Texture is an interesting characteristic of plants with which to work. There is a whole range of textures from coarse to fine that can be found in trees, shrubs and groundcovers. Since textures are created by the sizes of leaves and twigs, there is considerable variety. Texture is also affected by value. A grey-green foliage, for example, will often appear much finer than a bright green foliage with the same physical proportions. Because value and texture work so well together, the best textural contrasts are frequently those that use two or more distinct values of green with two or more distinctly different sizes of leaves. The combination of mugo pine and sea buckthorn shown in Figure 32 illustrates this point.

To make significant use of texture, a designer must develop a familiarity with plant materials. Strong contrasts of texture are quite interesting, as are subtle harmonies that can be developed through the juxtaposition of similar foliage types. The impact of texture in the landscape will lessen and disappear entirely as the distance from the observer to a group of plants increases. In such cases the impact of a composition which was highly textural when viewed close up becomes nothing more than a composition involving the silhouette of form and size when viewed beyond a certain distance. Because of the effect of distance on the perception of texture, it makes good sense to locate major textural groupings close to the more intensively used parts of the garden where the arrangement can always be appreciated.

Figure 32. Value and Texture

Woody Ornamentals

Figure 33. Planting Design

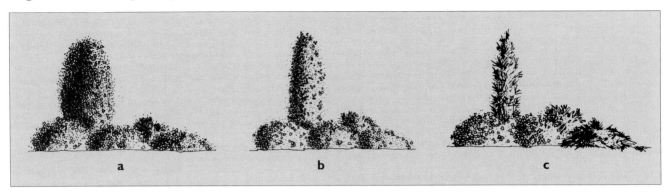

Principles of Composition

A designer has the opportunity to do quite a number of things with plants based on the degree of variation that exists because of plant form, size, color and texture. The following example illustrates some of the basic considerations that must be made when undertaking a problem in planting design.

In Figure 33a, the most common shrub form is that of the oval. One is an upright oval; the others are long in the horizontal dimension. The two forms can be combined to provide a very low-key type of composition, particularly if they are both deciduous and share a similar color. On the other hand, the impact of such a composition could be increased if the more upright form selected were changed to a very narrow columnar form, as in Figure 33b. If the new upright form chosen were changed to a conifer, along with one of the horizontal ovals (see Figure 33c), the impact of the composition would be further increased because of the marked contrast in foliage types.

It seems obvious from the example just given that compositions become more interesting as the number of contrasts involved increases, or as the magnitude of the contrasts increases. This is true up to a point., however, it must be realized that the term "composition" implies something that also has a strong feeling of unity. A planting scheme that simply involved many different contrasts would not meet the qualifications of composition on the basis of the variability introduced.

Use of a Dominant Element

Designers have several ways of achieving and retaining unity. In a simple planting scheme, it may be arrived at by the use of a dominant element. For example, if a plant group were to consist of a block of three silver-leaved plants and a block of three red-leaved plants, all of roughly the same size and shape (see Figure 34a), the arrangement would attract some attention, but it would hardly meet the requirements of composition. If, on the other hand, the plants in one group were twice as large as those in the other (see Figure 34b), dominance would be achieved through the larger elements, and unity (hence composition) would result.

The Value of Repetition

Repetition is another form of elemental relationship that is commonly used to achieve unity in planting design. For example, it does not require much in the way of imagination to envisage how the distribution of three or four plants of a single, spring-flowering shrub or tree throughout a small garden would tend to hold things together, at least during the spring season. The repetition of color, form, and texture, or any combination of these elements, is by far the most commonly used means for achieving unity.

Repetition can also be used in very subtle ways to achieve unity. The designer of the garden shown in Figure 35a has used two quite different plants to arrive at this goal and yet repetition has still been used to get there. In this

Figure 34. Achievement of Unity: Use of the Dominant Element

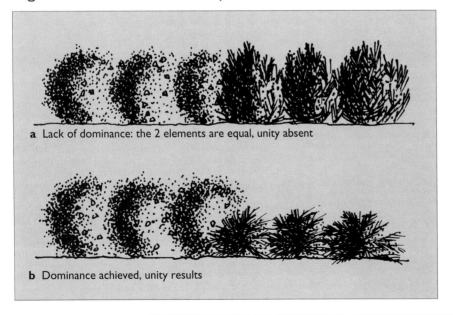

a Lack of dominance: the 2 elements are equal, unity absent

b Dominance achieved, unity results

case, composition has been achieved not by repetition of the same plant but by repetition of the same foliage color. A similar approach can be seen in the garden shown in Figure 35b; however, in this case it is a similarity in texture rather than color that has been used to achieve unity.

The Importance of Contrast

There is one other comment that must be made relative to the use of repetition in Figure 35a and that is, because of the great difference in size of the elements used (tree versus shrubs), a high level of contrast has also been introduced. Contrast, as was pointed out earlier, is the elemental relationship that functions to make a composition interesting.

In the foregoing examples, two important aspects of design aesthetics have been illustrated:
- The use of dominance or repetition to create unity
- The need for contrast to achieve interest.

In the three examples in Figure 36, it is not difficult to recognize the repetitive elements that have been used to achieve unity. Neither is it difficult to recognize the elements of contrast that are used to provide interest. In (a) the elements of contrast are two planes – one horizontal and one vertical. In (b) it is the positioning of the rectangular islands within the rectangular water body so that the watery space between varies in both size and shape. In (c) it is the orientation of the two curved

Figure 35. Unity in Planting Design

surfaces used in the architecture of the space. It is horizontal in the building, vertical in the arch.

Harmony as a Design Form

Both repetition and contrast have been referred to as forms of elemental relationship. One other form of elemental relationship that remains to be addressed is harmony. Harmony is a middle-of- the-range relationship, something that falls in between repetition at one end and extreme contrast or discord at the other.

Compositions that contain a high degree of harmony are less likely to be as lively and exciting as compositions in which some degree of contrast has been stressed. Neither will they be as monotonous and uninteresting as compositions that rely totally on exact repetition. This is not to say that compositions that rely on harmony are any less desirable than compositions that emphasize the relationship of contrast. People's preferences vary and design situations vary. J.O. Simonds illustrates this in his interesting book, *Landscape Architecture,* by pointing out what would happen if a designer chose to apply the same form of elemental relationship to the design of such disparate projects as:
- A fine restaurant
- A children's play space.

If the approach in both cases was composition to achieve harmony, in the first instance the scheme would likely have a very pleasing effect on the diners. However, in the second, it would likely put the children to sleep. By the same token, if a high degree of contrast were used in the two compositions, in the first case, users would likely end up with indigestion while the clientele of the second would likely become active, happy and quite vocal.

Figure 36. The Importance of Repetition and Contrast to Design

Woody Ornamentals

Planting Design in Garden Design

Because planting design and garden design are parts of the total landscape design package, it follows that a very strong relationship between the two must be developed right from the beginning of the conceptual design stage. The soft landscape must relate to the elements of the hard landscape. The lines, the forms, the materials and the structures of the hard landscape may call for the use of specific plant types and specific ways to use them. The more urban a landscape setting becomes, the more the elements of the hard landscape can be expected to exert their influence on the elements of the soft. Because of this, at least in North America, the man-made landscapes of the urban type will be much more structured than those found in the more rural areas.

Because of their size and setting, urban gardens are somewhat like fishbowls in that they turn inward on themselves. Space in the small garden is quite unlike that which exists on suburban and rural properties. In rural properties, the garden designer may, if the situation exists, extend the garden, in a visual sense, far beyond the boundaries of the property.

Because of space restrictions in the small garden, the planting approach generally recommended is group planting. This approach is aptly named because it involves looking at the conceptual plan for the garden and then making decisions as to where and what plant groups are needed to augment it. When these decisions have been made, the plan can then be checked in its entirety to see how well the combination of the various plant groups meets the overall need for unity and interest. Finally, the group plantings can be linked by groundcover, herbaceous perennials or elements of the hard landscape to become part of the total garden design.

The small garden shown in Figure 37 uses these principles. Small groups of shrubs, some of them identical, have been distributed to unify the garden. The shrub groups have also been linked with masses of groundcover to increase the level of interest and tie the elements together.

On large sites, the planting design approach will be quite different. This is partly because the visual limits are not as well defined as they are on the smaller urban property and partly because plant masses which form the structure of the landscape are not necessarily linked and physically continuous.

The Japanese have an interesting way of looking at garden design, at both scales. Their approach to the smaller urban garden might be to develop a 'landscape within a garden', that is, to miniaturize an actual landscape within an enclosed space. One interesting example of this is the historic Zen garden **Ryoanji** which uses large stones to represent islands and mountains, and carefully raked sand to represent the ocean. For the larger property, which we would describe as suburban or rural, the objective most

Figure 37. Planting and Garden Design for an Urban Property

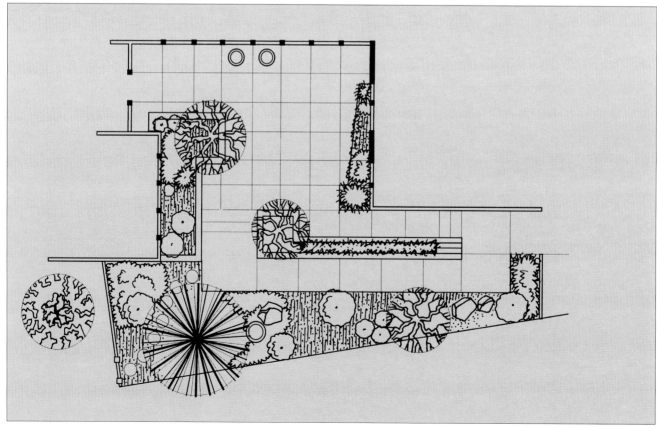

frequently employed by the Japanese is to design a 'garden within the existing landscape'. The Japanese approach to garden design on small properties would never suit most North Americans, but their approach to garden design on larger properties, that is, 'garden within the existing landscape,' most certainly would.

'Garden within the existing landscape' starts with evaluation of what is to be seen. Some property owners do this prior to siting the house, so that it might relate to the best views, to existing trees, to topographical features, and so on. When this is done, additional plantings can be used to relate the house to the important aspects of the natural landscape. The structuring of the plantings on the large site is not as obvious as it is on the small site; nevertheless, planting design must be just as carefully conceived so that it might exist in harmony with the overall natural landscape. For example, it is important that the most frequently chosen tree type bear some similarity with the existing ones. When such trees have at least a similar size and form as those that exist, unity is achieved.

The large landscape also lends itself more readily to the use of trees as accent materials. Trees with outstanding characteristics like flower or foliage color are frequently used to draw attention to some part of the site. Such accents are most effective with good backgrounds. As such, they are generally used singly or in small groups in front of the plant masses that define the landscape spaces.

Because planting design for the large site is essentially mass planting, shrub selection for such sites is not nearly as critical as it would be on the small site. Many shrubs that would be unacceptable as single specimens or in small groups can become quite desirable on a large site when used in mass. In such situations any shortcoming a plant might have had as a single specimen is completely masked when the plant is closely associated with others of the same type.

In summary, the approach to planting design for both large and small properties, while essentially different, still involves quite a lot of common ground. Plants must always suit the environment into which they have been placed and the aesthetic principles of design are universal. Any difference taken in design approach will likely occur because of differences in scale and because of differences in the environments to which the projects are assigned.

Figure 38. A Typical Planting Arrangement for a Large Site

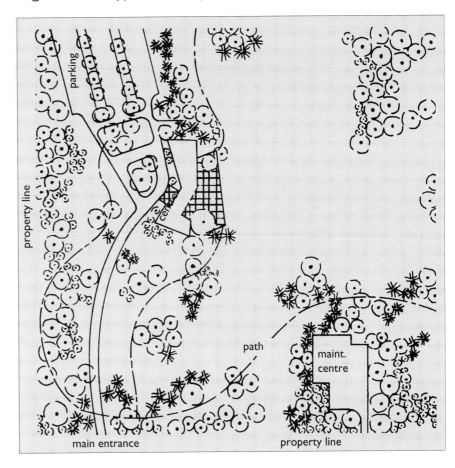

References

Simonds, J. O. 1961. *Landscape Architecture.* F. W. Dodge: New York.

Plant Descriptions

How to find a specific plant or plant type

Plant Descriptions

One of the major problems encountered in putting together a book of this sort is how to overcome the difficulties associated with botanical names with which many of its users are unfamiliar. In this, the revised edition, we hope that botanical names will no longer be a problem, not that we have abandoned botanical nomenclature and the international rules for its usage, but rather because we have added an indexing system that cross-references the common name of a plant to its generic equivalent. This, we hope, will make the book much easier to use.

The subject matter of the Plant Description section and the Reference Charts is totally descriptive. In the Plant Description section, you will find detailed information about all of the plants, a photograph, and a silhouette to help you visualize the size and shape of this plant in your landscape. This section is organized alphabetically according to botanical name.

A glossary of botanical and horticultural terms is included on page 207 and the cross-referencing index begins on page 209.

Reference Charts

The Reference Charts have been included to make use of the book as easy and convenient as possible. If a user requires a quick check on the characteristics of a particular type of plant, this section of the book will be helpful.

Separate charts have been developed for coniferous and deciduous materials, as well as for groundcovers and for vines and climbers. To make things as useful as possible, the charts for coniferous and deciduous materials each have two sub-sections – one for shrubs and the other for trees.

The characteristics referred to in the charts are those that will commonly affect choice, such as size, form, texture, color of foliage and flowers, and special features such as those which might well be considered as limiting factors.

For convenience, all plants are listed alphabetically by common name and each entry includes the page number where further descriptive information, including botanical name, is to be found.

Tree Form

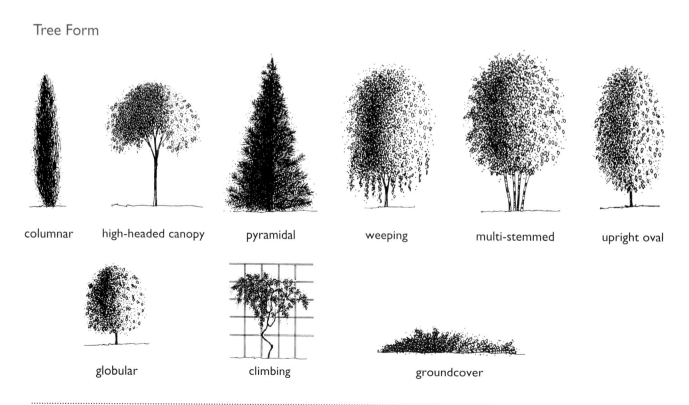

columnar   high-headed canopy   pyramidal   weeping   multi-stemmed   upright oval

globular   climbing   groundcover

Woody Ornamentals

Abies

Fir

The tree types are all medium-sized, narrowly pyramidal conifers. Bark is smooth and covered with conspicuous resin blisters. Needles are soft and flattened. In the species hardy here, the needles are arranged in one plane, though in some they will turn upwards rather than lie flat. They are fragrant, soft and flexible; tips are blunt or notched. Fir needles, unlike those of spruce, do not have cushions or pegs; hence needle scars are not raised above the branches.

The one hardy dwarf fir is a variant of the tree species. It does not resemble parental material in form or size, but foliage characteristics are quite typical of the genus.

Cones of all species are erect and candle-like. When the seed crop is mature, the cone scales drop, exposing an erect, spike-like central stalk (rachis) which persists.

Native species are found only in areas of highest rainfall – the foothills, mountains and northern regions of the provinces. All are shallow-rooted trees and should be grown only in well-mulched, sheltered locations. In spite of their preference for a sheltered habitat, one species, *Abies lasiocarpa*, is commonly found growing at the tree line in the Rockies where the harsh environment severely restricts its growth giving rise to dwarfed and deformed specimens collectively referred to as "krummholz".

Abies balsamea
Balsam Fir

Size	10 m.
Form	Dense, narrow pyramid.
Foliage	Dark green, soft, flat with notched tips with two white lines beneath.

Seed Cones 7.5 cm, erect, violet purple.

The balsam fir prefers a cool, moist, sheltered site. It retains its symmetrical form for at least 15 years but becomes loose and unkempt as it gets older. This condition is accelerated under dryland conditions.

Abies balsamea

Cultivars of *Abies balsamea*

'Nana' This small, spreading cultivar lacks a vertical growing point; hence it grows only horizontally. Growth tends to radiate from the center of the plant. One of the better dwarf conifers for shady areas.

Abies balsamea 'Nana'

Abies concolor

Abies concolor
White Fir, Colorado Fir

Size	8 m.
Form	Dense, narrow pyramid.
Foliage	Silvery green; needles flat, but turned up at the tips with white stomatal lines on both surfaces; needles are rounded at the apex.
Bark	Silvery-grey.

Seed Cones Green or purple, cylindric, 7–12 cm.

At this latitude *A. concolor* prefers sheltered locations but may tolerate drier conditions than the other species. In some parts of the region it can be expected to be of borderline hardiness but when the mesoclimate is favorable, full hardiness can be expected.

Abies lasiocarpa

Abies lasiocarpa
Subalpine Fir

Size	8 m.
Form	Narrow pyramid.
Foliage	Dark, bright green.

Seed Cones 8 cm, deep purple.

Moist, sheltered locations are preferred. This is an attractive tree for 15–20 years, but it becomes quite open with age, often losing its lower branches.

Abies sibirica

Abies sibirica
Siberian Fir

Size	10 m.
Form	Narrow pyramid.
Foliage	Bright, shiny green, grooved above, tip entire. Needles point forward towards the tips of branches. The undersides of needles have two greyish lines.
Bark	Smooth, greyish.

Seed Cones 5–8 cm, cylindric, blue.

This is a beautiful, fresh-looking, hardy specimen tree. It is inclined to start growth early, so could be injured occasionally by late frosts. It is occasionally subject to aphid infestations.

Acanthopanax

Pricklyspine

Acanthopanax is a group of large coarse-textured deciduous shrubs native to eastern Asia and the Himalayas. In spite of the common name the spines are short and few in number. As an ornamental, *Acanthopanax* has very few outstanding characteristics. The leaves do not color up in the fall and its flowers and fruit, while interesting, are not showy. Leaves are coarse, palmately compound and alternate. The inflorescence is a short-stalked umbel and the fruits are borne in such tight clusters that they appear to be compound. The fruit, which is generally abundant, is quite edible and, when preserved, is said to make a good substitute for blueberry jam. The plant has no particular site preferences but does very well in shade. It makes good background material.

Acanthopanax sessiliflorus
Pricklyspine

Size	3 m.
Form	Globular, becoming leggy with age.
Texture	Coarse.
Foliage	Light green.
Flowers	Purplish.
Fruit	Because of short pedicels and the globose character of the inflorescence, the fruit cluster resembles a giant black raspberry fully 3–4 cm across.

This shrub is adapted to most soil and environmental conditions. It is large and can provide suitable background material in a shrub mass. It is not widely used.

Acanthopanax sessiliflorus

Acer

Maple

Large, shallow-rooted deciduous trees, large shrubs or small shrub-like trees. Leaves are opposite, palmately-lobed and simple in all species except *A. negundo*, which has a compound leaf with three to five leaflets. Flowers are imperfect and both male and female flowers can be either on the same tree or on separate trees, depending on the species. Fruit is dry and two-winged, splitting when ripe into two halves, each with a wing.

Woody Ornamentals

Acer ginnala

Acer ginnala
Amur Maple

Size	4–5 m.
Form	Low-headed, globe-shaped tree or large, upright-spreading shrub.
Habit	Decurrent.
Canopy	Dense.
Texture	Medium.
Foliage	Green, three-lobed leaves often with red veins and petioles. The terminal lobe is much longer than the other two.
Flowers	Small greenish yellow in clusters.
Bark	Smooth, greyish, patterned very much like snake skin.
Fruit	Each wing of the two-winged fruit resembles a miniature lobster claw. The margins of the fruit are bright red.

Autumn color of the foliage is outstanding. Color from the fruit is also conspicuous and ornamental. The fruit colors up in August, the foliage in September. Autumn foliage color is best on plants growing in sunny locations.

Acer ginnala

Cultivars of *Acer ginnala*

var. *seminowii* A shrubby plant with smaller leaves which are occasionally five-lobed.

Acer glabrum var. *douglasii*

Acer glabrum var. *douglasii*
Rocky Mountain Maple

Size	4 m.
Form	Low-headed, upright.
Habit	Decurrent, short trunk with many ascending branches.
Canopy	Open.
Texture	Coarse.
Foliage	Green leaves on long, red petioles, three-lobed to indistinctly five-lobed. Autumn color is yellow to dull red.
Flowers	Greenish yellow, imperfect. In this species the male and female flowers are found on separate trees.
Bark	Dark reddish brown.
Fruit	Rose colored when ripe.

This is an attractive, small, flowering tree for shady moist woodland habitats. Flowers are borne in clusters on short stalks. This plant is not hardy in all areas.

Acer negundo
Manitoba Maple, Box-Elder

Size	12–15 m.
Form	High-headed, upright-oval.
Habit	Decurrent.
Canopy	Open.
Texture	Coarse.
Foliage	Yellowish green; leaves are compound with three to five leaflets.
Flowers	Imperfect/dioecious.
Bark	Dark brown, thick, furrowed. The trunks of older trees frequently become deformed with large rotund "swellings". These contain clusters of internally located trace buds that become active in times of stress and are responsible for the emergence of "water sprouts".

Acer negundo

This is a weedy tree of doubtful value to the landscape. It has two particularly bad habits: 1. the germination and growth of unwanted seedlings and 2. an inherent ability to attract plant lice or aphids, which produce a sticky substance called honey-dew that literally drips from the tree canopy and damages the surface of anything beneath it.

Despite these shortcomings, older trees often make good climbing structures for children and great locations for tree houses. The variegated form 'Variegatum' that is popular in the north-western U.S.A. has not proven hardy here.

Acer negundo

Cultivars of *Acer negundo*

'Baron' A male clone, selected to avoid the objectionable habit of producing seedlings wherever seed may fall.

Acer platanoides
Norway Maple

Size	8 m.
Form	High-headed, upright-oval.
Habit	Decurrent.
Canopy	Dense.
Texture	Coarse.
Flowers	Bright yellow in small spreading panicles.
Foliage	Green, large, three-lobed with wide sinuses. Petioles will exude a milky juice when leaves are removed from the stems.

Acer platanoides

The Norway maple is worth trying in some of the more favorable environments in the region. Some success has been experienced in the larger cities where the mesoclimate is favorable.

Woody Ornamentals

Cultivars of *Acer platanoides*

Two color forms of this species are seen occasionally. Both are subject to winter injury in the early going but seem to have the ability to acclimate with time.

'Crimson King' Similar in form to the previous cultivar but with red foliage.

'Schwedleri' A tree with a dense compact head of deep-purple foliage.

Acer platanoides 'Crimson King'

Acer saccharinum
Silver Maple

Size	16–20 m.
Form	High-headed, upright oval.
Habit	Decurrent.
Canopy	Open.
Texture	Coarse.
Foliage	Leaves are deeply cleft, five-lobed, light-green above, and silvery beneath.
Flowers	Small, deep-red, produced early, before the leaves, in short, almost stalkless clusters.
Bark	Smooth, distinctive light-grey becoming rough with thin, loose, vertical flakes as it gets older.

Hardiness may be a problem in rural areas beyond the heat island of the city. Leaf characteristics contribute greatly to its ornamental value. Flower buds and flowers, though small, are quite conspicuous. Autumn color in the region is a clear yellow.

Acer saccharinum

Cultivars of *Acer saccharinum*

'Northline' A Morden introduction, said to be a slower growing tree than the species.

Acer saccharinum

Acer saccharum
Sugar Maple

Size	10 m.
Form	High-headed, upright-oval.
Habit	Decurrent.
Canopy	Dense.
Texture	Coarse.
Foliage	Green, leaves three to five lobed with broad, shallow sinuses.
Bark	Dark grey in firm vertical strips.

The sugar maple is not widely grown in the region because its hardiness has been in question. Nevertheless good specimens of this tree are to be found in the prairie provinces and it is certainly worthy of trial. Sheltered locations must be selected. Autumn color is bright red.

Acer saccharum

Acer spicatum
Mountain Maple

Size	3–5 m.
Form	Low-headed, upright, shrub-like.
Habit	Decurrent.
Canopy	Open.
Texture	Medium.
Foliage	Green, three-lobed or loosely five-lobed leaves with long reddish petioles.
Flowers	Creamy, very showy in upright clusters which appear after the leaves are full grown.
Bark	Smooth, reddish.
Fruit	Red, showy.

This is an interesting, small, shrubby tree for sheltered sites. Although it is not hardy in all areas and prefers acid soils, it is worthy of trial.

Acer spicatum

Acer tataricum
Tartarian Maple

Size	5 m.
Form	Low-headed, upright-oval.
Habit	Decurrent.
Canopy	Open.
Texture	Medium.
Foliage	Green, leaves ovate with bright red petioles; fall color, yellow or orange.
Flowers	Fragrant.
Fruit	Showy red fruit in late summer.

This maple prefers a rich, deep soil in open, sunny locations. It is not quite as hardy as the Amur maple and not commonly seen. Its fall color is outstanding.

Acer tataricum

Woody Ornamentals

Aegopodium

Goutweed, Goatsfoot

These perennial herbs have ternately compound leaves; leaves are variegated, white and green. Inflorescence is a compound umbel with lacy white flowers. This plant is a European native that has escaped and established itself in North America. As a groundcover it does extremely well in shady, moist environments. It is considered a nuisance plant by most people; however, it can be useful in rare cases when its shortcomings have been recognized.

Aegopodium podagraria

Aegopodium podagraria
Bishop's Goutweed

Size	0.4 m.
Form	Upright-spreading, not woody.
Texture	Medium.
Foliage	Variegated, white and green.
Flowers	White.

This is used as a groundcover for shade and semi- shade. However, it is a very invasive plant and therefore should not be used with less competitive plants. Woody plants can generally compete. The tops of *Aegopodium* are killed by fall frost, but the roots are winter hardy.

Aesculus

Buckeye, Horse Chestnut

Attractive medium to large flowering trees with stout branches and very dense compact heads. Leaves are coarse-textured, palmately compound, with five to seven leaflets. Winter buds are large and sticky in the horse chestnut, but not sticky in the Ohio buckeye. Both species are noted for their flowering habit and are somewhat rare in the region. The Ohio buckeye is fully hardy in most areas and the horse chestnut will survive in the larger cities where the mesoclimate is favorable.

Aesculus glabra
Ohio Buckeye

Size	6–8 m.
Form	Low-headed, round to upright-oval.
Habit	Decurrent.
Canopy	Dense.
Texture	Coarse.
Foliage	Leaves are green, large, palmately-compound. There are usually seven leaflets, each 10–15 cm long with long, pointed tips and wedge-shaped bases. Winter buds not sticky.
Flowers	White or pale yellow in upright panicles, in late June.
Bark	Grey, patterned, corky.
Fruit	A brown nut covered with a green, prickly husk; not edible.

This tree is tolerant of the usual soil and environmental conditions to be found in urban communities and sheltered farmsteads. It is fully hardy. The light-orange autumn color is good but not outstanding.

These trees have a tap root hence larger specimens may be difficult to transplant successfully. Containerized stock is recommended for specimens over 2 m in height. Because of their density, trees can create strong rain shadows. When used as lawn specimens, fruit can crate minor maintenance problems.

Aesculus glabra

Aesculus glabra

Aesculus hippocastanum
Horse Chestnut

Size	10 m.
Form	Low-headed, globular.
Habit	Decurrent.
Canopy	Dense.
Texture	Coarse.
Foliage	Leaves are green and palmately compound with seven leaflets per leaf. Leaflets are obovate with wedge-shaped bases and abruptly sharp-pointed tips. Winter buds are sticky.
Flowers	Upright-panicles, very showy.
Fruit	A large brown nut covered by a prickly husk; not edible.

This tree is suited to average soil conditions in favorable mesoclimates. Hardiness is borderline. It is definitely not hardy enough for the northern parts of the provinces.

Aesculus hippocastanum

Alnus
Alder

The alders are large, coarse shrubs or small trees with a fruiting habit similar to that of birch, that is, flowers are borne in catkins with both male and female types on the same tree. Buds are alternate and borne on prominent short stalks. The leaves are oval and toothed with two sizes of teeth. Like its relative, the birch, the bark is usually covered with lenticels, but it is silvery grey or brown and not papery. Alders are shallow-rooted and are commonly found growing in wet places.

Alnus rugosa

Alnus rugosa
Red Alder

Size	5 m.
Form	Upright-spreading.
Habit	Decurrent.
Canopy	Open.
Texture	Coarse.
Foliage	Green.
Bark	Smooth, light grey, glossy.

A fast-growing, hardy small tree for sunny, wet locations.

Alnus tenuifolia

Alnus tenuifolia
Mountain Alder

Size	5 m.
Form	Low-headed, upright- spreading.
Habit	Decurrent.
Canopy	Open.
Texture	Coarse.
Foliage	Dark-green.
Bark	Smooth, grey to brown.
Fruit	A small, oval, cone-like structure that persists long after the seed has been shed.

This plant prefers wet sites in open sunny locations. It is not widely used as a landscape plant and is best suited to wildland situations. It is subject to attack by leaf miners.

Amelanchier

Saskatoon

Only one species is indigenous to the region; however, there is wide variation within the species. One of the more unusual fruiting kinds is a white-berried form that is just as sweet if not sweeter and juicier than the type.

The saskatoon is generally a tall upright-spreading shrub. It is a good flowering plant that blossoms in the spring before the leaves have reached full size. Leaves are borne alternately, are approximately 5 cm long, and are oval shaped, serrated at the apex. Buds are sharp-pointed and slender.

The saskatoon is grown chiefly for its fruit, though in recent years people have begun to recognize the ornamental value that lies chiefly in its flowering habit. In some instances, plants have also been selected because of their autumn coloration and form. Flowers are white and borne in terminal racemes, appearing about mid-May. Fruit is a berry-like pome, with a persistent calyx. It is generally 1–1.5 cm in diameter, deep blue, juicy and sweet when ripe, but contains little pectin.

Amelanchier alnifolia
Saskatoon

Size	4 m.
Form	Upright-spreading, leggy.
Texture	Medium.
Foliage	Dark green.
Flowers	White, showy, and produced before the leaves.
Fruit	Red, ripening to blue-purple, sometimes white.

This plant has few specific requirements. It grows well on a wide variety of soils and seems to do well in sun or partial shade. Plants growing in deep shade usually become very long and spindly with few leaves. In residential landscapes, it can be used like any large flowering shrub. Because of its legginess, smaller plants can be planted in close association with it. In recent years, commercial fruit growers have reported losses due to *Cytospora* canker infection.

Amelanchier alnifolia 'Pembina'

Cultivars of *Amelanchier alnifolia*

'Altaglow' Columnar to pyramidal in form to 7 m. Fruit is white, sweet, and sparsely produced. It does not sucker freely. Fall color is orange to red.

'Honeywood' A late fruiting cultivar that produces heavily when young. Large trusses of fruit, up to 15 per truss have been reported. Height to 5 m.

'Northline' Upright-spreading shrub to 5 m, very productive. It suckers freely.

'Pembina' Upright-spreading shrub to 5 m. It produces few suckers. Fruit is large, fleshy, and full flavored.

'Smoky' Upright-spreading shrub to 5 m. Fruit is unusually sweet, but mild flavored. The plant suckers freely.

Amelanchier alnifolia 'Smoky'

'Thiessen' A columnar plant to 5 m with fruit 17 mm in diameter. It is not reliably hardy in northern parts of the region. Uneven fruit ripening is also a characteristic of this cultivar.

Anemone

Anemone

Anemones are erect perennial herbs; the basal leaves are generally dissected. The flowers are apetalous but the sepals are quite showy. The achenes produced are compressed, one-seeded and either pubescent or woolly.

Because of their ability to produce offsets some species make effective flowering groundcovers for both full sun and partial shade. The species described below is by far the best for groundcover purposes.

Anemone sylvestris

Anemone sylvestris
Windflower Anemone

Size	0.3 m.
Form	Low, spreading plant.
Texture	Fine.
Foliage	Dark green, finely-cut leaves.
Flowers	White, 5 cm in diameter.

This is one of the better groundcovers for a wide variety of locations. Planted on 15-cm centers, this plant will cover the ground in one season and overcome competition from weeds. It is a European species and should not be mistaken for the native species, *Anemone canadensis,* which is definitely inferior.

Antennaria

Pussytoes

Mat-forming perennial herbs. Leaves are woolly, usually grey-green. Mother plants and offsets have dense rosettes of leaves and spread by short stolons. The flowers are imperfect/dioecious and borne terminally on upright stems in many-flowered heads. Because of their dense mat-forming growth habit, they are good groundcovers. They prefer sunny locations.

Antennaria plantaginifolia var. *ambigens*
Pussytoes

Size	0.1 m.
Form	Low mat-forming perennial.
Texture	Medium.
Foliage	Silvery, woolly.

This is a good groundcover for full sun. The silvery foliage adds to its value. Flowers are conspicuous but not ornamental.

Antennaria plantaginifolia var. *ambigens*

Arctostaphylos

Bearberry, Kinnikinik

This genus includes two native plants. Both are low, spreading groundcovers bearing fruit that is an important food for wildlife in the environments they occupy. As groundcovers, both have much to commend them. Both prefer sunny environments and do well on light sandy or gravelly soils. Both species have small urn-shaped flowers, but here the similarities end: one is a broad-leaved evergreen with long, creeping stems and bright glossy leaves; the other deciduous with interesting strongly-veined foliage that turns a striking bright scarlet in the fall.

Arctostaphylos rubra
Alpine Bearberry

Size	0.15 m.
Form	Low, ground-hugging plant.
Texture	Medium.
Foliage	Dull green with a strong network of veins which can be noticed on both sides of the leaf.
Fruit	Dark, wine-red, juicy berry.

This plant prefers sunny locations. In the wild it grows on thin, gravelly soils in areas of high rainfall. It is extremely colorful in the fall season when the leaves turn a bright scarlet.

Arctostaphylos rubra

Woody Ornamentals

Arctostaphylos uva-ursi

Arctostaphylos uva-ursi
Kinnikinik

Size	0.1 m.
Form	Prostrate, with long primary stems.
Texture	Medium.
Foliage	Glossy green, thick.
Flowers	Pinkish white.
Fruit	Scarlet, dry, 1-cm diameter.

This native groundcover grows well in light soils in sunny, open situations. The leaves are retained the year round, so it is a broad-leaved evergreen. Native stands can be found in sandy soils throughout the north-central and northern parts of the region. It can be readily propagated from softwood cuttings; transplants from the wild can also be used but should be heavily cut back prior to transplanting.

Cultivars of *Arctostaphylos uva-ursi*

'Vancouver Jade' This plant is said to spread more quickly than the wild form. Flowers are dark pink and fragrant.

Berberis

Barberry

The hardy barberries are spiny, stiff-stemmed deciduous shrubs. Their ornamental value lies in their foliage color and fruit in most cases. The fruit of the barberry is a somewhat elongate berry which in most cases is red. Flowers are yellow and the sepals are petal-like.

Because some species have been identified as the alternate host of stem rust of wheat, they may be on the restricted list in some provinces. Information regarding restricted species can be obtained from provincial departments of agriculture.

Berberis koreana
Korean Barberry

Size	1 m.
Form	Upright-spreading.
Texture	Medium.
Foliage	Green.
Flowers	Small, golden-yellow.
Fruit	Bright red, sub-globose in clusters.

The Korean barberry prefers a sunny location. It retains its colorful fruit over a long period.

Berberis koreana

Berberis poiretii
Poiret's Barberry

Size	1.5 m.
Form	Plant with elegant drooping branches.
Texture	Medium.
Foliage	Brilliant fall color.
Flowers	Pale yellow produced in abundance.
Fruit	Slender elongate bright red berries.

Likely the showiest of the hardy barberries.

Berberis poiretii

Berberis × 'Sheridan Red'
Sheridan Red Barberry

Size	1 m.
Form	Upright-spreading.
Texture	Medium.
Foliage	Deep reddish purple.
Flowers	Yellow.
Fruit	Rare

This hybrid has the common barberry, *B. vulgaris* in its parentage and for this reason the hybrid is no longer carried by western Canadian nurseries. Nevertheless it can still be seen in many parts of Alberta. It is an attractive plant because of the deep purple foliage and because at least some of its stems tend to branch and avoid the stiffness that most barberries display. However it is not an easy plant to maintain because many of its very spiny stems have the unfortunate habit of winterkilling most years.

Berberis × 'Sheridan Red'

Woody Ornamentals

Bergenia

Saxifrage

These low-growing, coarse-textured, perennials from Siberia are very useful landscape plants. Leaves are large, dull green, sometimes glossy, and are retained over winter. Generally they become tinged with red as the cold weather of autumn approaches. Colonies spread slowly from large fleshy stolons. The plants are of easy culture, and will grow in all but the lightest soils, and seem to do well under a variety of light conditions. Flowers are quite showy and are produced well above the leaves in late spring.

Bergenia cordifolia

Bergenia cordifolia
Large-leaf Saxifrage, Siberian-Hyacinth

Size	0.3 m.
Form	Coarse-textured groundcover.
Texture	Very coarse.
Foliage	Large, circular, green leaves that are quite purplish on their undersides in spring and fall but become green as the soil warms up.
Flowers	Pink, showy in spike-like racemes.

This is an attractive, hardy herbaceous plant with very coarse leaves that provide excellent textural contrasts with finer textured subjects. The leaves are persistent. When low temperatures occur at the end of the growing season, the leaves respond by reducing free water in the tissues. This causes them to lose their ability to stand upright and they will lie flat on the ground over winter. They remain so until soil temperatures rise, enabling the plant to take up water. In response to these conditions, the leaves quickly resume an upright position.

Betula

Birch

This genus is made up of attractive trees and shrub-like trees, many of which rank high as specimens. The bark is aromatic, white or red, papery or leathery, with narrow horizontal markings (lenticels). Leaves are alternate and simple; flowers are imperfect/monoecious. The flowers are catkins; those of the male are formed in autumn and are elongate; female catkins are smaller, oblong or cylindrical.

Birch trees are not particularly long-lived due primarily to the effects of wind and prolonged periods of drought. In the drier parts of the region, die-back from the top of the tree is a common occurrence. When birches become unthrifty, quite frequently the bronzed birch borer, *Agrilus anxius*, enters the picture and administers the *coup de grace*. This small insect lays eggs on dead or damaged branches, which generally appear first at the top of the tree. The larvae tunnel their way downwards just under the bark, leaving evidence of their spiral tunnels. Adults emerge in the spring to deposit their eggs at the lower end of the dead branches. Cutting out dead limbs a little below the lowest point at which damage is noticed appears to be the best method for control of this pest. Prunings should be burned.

Birches do not adapt well to fall planting and do not tolerate a lot of pruning. They will "bleed" profusely from cuts made to leafless stems in the spring. "Bleeding" can be avoided by delaying any necessary pruning until after the leaves have reached full size.

Betula albo-sinensis
Chinese Paper Birch

Size	to 15 m.
Form	Upright oval.
Habit	Decurrent.
Canopy	Dense, closed.
Texture	Coarse.
Foliage	Green.
Bark	Pink or red exfoliating bark with a glaucous "bloom".

An attractive tree grown for its interesting bark color.

Betula albo-sinensis

Betula davurica
Dahurian Birch

Size	10 m.
Form	Upright-oval.
Habit	Decurrent.
Canopy	Closed, open in older trees.
Texture	Medium.
Foliage	Green.
Bark	Interesting because of its tendency to exfoliate and yet be retained. Because of the interesting bark habit, this is an outstanding specimen tree.

Betula davurica

Woody Ornamentals

Betula fontinalis

Water Birch

Size	6 m.
Form	Low-headed, globular.
Habit	Decurrent, frequently multi-stemmed.
Canopy	Open.
Texture	Medium.
Foliage	Green.
Bark	Dark shiny brown; does not peel.

This tree prefers moist, open sites. It can be readily propagated from old stumps if young trees are not available. Leaves turn a dull yellow in the autumn. It is commonly found along watercourses in southwestern Alberta and central Saskatchewan.

Betula fontinalis

Betula nigra

River Birch

Size	5 m.
Form	Upright-oval.
Habit	Decurrent.
Canopy	Dense, closed.
Texture	Medium.
Foliage	Green, autumn color is golden.
Bark	White, with a strong tendency to peel in horizontal strips and tending to darken with age.

This tree is interesting, largely because of its very ornamental bark

Betula nigra

Betula papyrifera

Paper Birch

Size	10 m.
Form	Upright-oval.
Habit	Decurrent.
Canopy	Open or closed.
Texture	Medium.
Foliage	Green, turning bright yellow in the fall.
Bark	Ornamental, white.

The paper birch adapts well to most sites but is very susceptible to prolonged drought. In recent years it has been attacked by a leaf miner which has spoiled the appearance of the foliage and caused much leaf drop from mid-summer onward. The use of the systemic insecticide dimethoate can

Betula papyrifera

give some measure of control when used at high concentrations as a soil drench. Winter color of this tree is very effective when viewed against a contrasting background.

Cultivars of *Betula papyrifera*

'Chickadee' An attractive, narrowly-upright tree with exceptionally white bark.

Betula papyrifera 'Chickadee'

Betula pendula
European Birch

Size	15 m.
Form	Low-headed, upright-oval.
Habit	Decurrent.
Canopy	Open.
Texture	Fine.
Foliage	Green.
Bark	Bark is white and papery. Black, corky, rough spots forming a more or less diamond shaped pattern are common on the lower trunks of older trees.

Betula pendula

This tree prefers a sandy soil, provided that moisture is not in short supply. The youngest shoots are a distinguishing feature. They are long and thin and arch downwards. Foliage turns bright-yellow in the fall.

Cultivars of *Betula pendula*

'Fastigiata' This cultivar is a small, attractive, columnar-tree with good, white bark and a lot of fastigiate growth. It has a tendency to become upright oval with age.

'Gracilis' (cutleaf weeping birch) This is very common throughout the region. It has a very pronounced weeping habit with long, thin, flexible branchlets. It takes up a lot of space at ground level. Leaves of the cultivar are deeply cut and are quite lacy.

Betula pendula 'Gracilis'

'Youngii' (Young's weeping birch) This is a small tree with an almost grotesque weeping or sprawling habit. It is more of an oddity than a useful landscape element.

Woody Ornamentals

Caragana

Caragana, Peashrub

This group of plants belongs to the pea family *(Fabaceae)*. Introduced to western Canada in the 1880s, it enjoyed considerable popularity in the early years as a hedge and windbreak plant. Today, even the small-statured species are rarely encountered in urban situations, but *C. arborescens* is still used in field shelterbelts. Leaves are compound, green or yellowish-green with no terminal leaflet. The size and shape of the leaflets vary with the species. All species are armed with spines or prickles . All are suited to dry situations and will thrive under the low rainfall conditions of most areas without supplementary water.

Caragana arborescens

Caragana arborescens
Common Caragana

Size	To 4 m.
Form	Upright to upright-spreading.
Texture	Fine.
Foliage	Green.
Flowers	Yellow.

This plant is of little ornamental value but can be useful as a shelterbelt plant in rural areas.

Cultivars of *Caragana arborescens*

'Lorbergii' (fern-leaved caragana) A very fine-foliaged mutant of the common caragana with yellow flowers. Foliage resembles that of the asparagus fern.

'Pendula' (weeping caragana) Branches are heavy, distorted and pendulous. The foliage is similar to that of the species. Commonly used as a graft on a 1 m standard of common or Sutherland caragana. On its own roots it can form a dense woody groundcover.

'Plume' One of the more ornamental members of the genus. It is a very fine-textured plant with loose billowing masses of foliage on pendulous stems. Height is 1.5 m.

'Sutherland' A tall, narrowly upright plant with many vertical non-branching stems. It is frequently used as a standard on which to graft weeping forms.

'Tidy' A form of the fern-leaved caragana with a straight, upright- spreading growth habit.

'Walker' A very fine-leaved form of weeping caragana. It is used on its own roots as a sprawling plant or grafted on a standard of common or Sutherland caragana.

Caragana arborescens 'Sutherland'

Caragana frutex
Russian Caragana

Size	2 m.
Form	Upright-spreading.
Texture	Fine.
Foliage	Dark green, four leaflets per leaf
Flowers	Yellow.

This caragana is well suited to all but the wettest sites. The species is of little ornamental value, but is much more attractive than *C. arborescens*.

Caragana frutex 'Globosa'

Cultivars of *Caragana frutex*

'Globosa' A dense globular-plant with dark-green leaves, it is closed to the base and grows to a height of about 1 m. This plant has juvenile characteristics; hence it does not produce flowers or fruit. It is frequently used as a hedge plant in some of the drier parts of the region.

Caragana jubata
Shagspine Caragana

Size	1 m.
Form	Upright.
Texture	Fine.
Foliage	Sparse, green, compound-leaves.
Flowers	White.

This is an oddity and quite unlike the usual deciduous shrub in that it consists solely of a few thick upright stems that are covered with prickles, stipules and fine woolly hair. It may make a suitable subject for a rockery but does not appear to lend itself to widespread use elsewhere.

Caragana jubata

Caragana pygmaea
Pygmy Caragana

Size	0.75 m.
Form	Compact-mound, closed to base.
Texture	Fine.
Foliage	Yellowish green. very fine textured leaflets.
Flowers	Bright yellow.

Sunny, dryland sites are preferred. It is a dense spiny compact shrub well suited for use as a low barrier hedge but of little ornamental value otherwise.

Caragana pygmaea

Woody Ornamentals

Celastrus
Bittersweet

These woody climbers wind themselves around posts or other support structures. Fruiting plants produce attractive branches for winter landscape effect and for winter bouquets.

Celastrus scandens

Celastrus scandens
American Bittersweet

Size	2 m.
Form	Climber.
Texture	Coarse.
Foliage	Ovate to ovate-oblong green leaves 5–10 cm. Autumn foliage is a striking bright yellow.
Flowers	Imperfect/dioecious. Flowers are borne terminally in 5–10 cm panicles.
Fruit	Clusters of 0.8-cm orange capsules which expose a scarlet seed covering (aril) when ripe.

A well-drained sunny location is essential; both male and female plants are required to assure production of the attractive persistent fruit. The American bittersweet may not be hardy in the more northerly parts of the region.

Celtis
Hackberry

The hackberry is not frequently encountered, but reasonably mature stands at Regina and parts of southern Manitoba suggest that it should be tried over a wider area. It is a medium to large tree with ascending arching branches and a rounded head. The flowers of the tree are borne singly in the axils of the leaves. Trees are either polygamous or monoecious. The flowers are not attractive but the stigmatic surface of pistillate ones has an unusual caterpillar-like appearance. This is visible about the time the leaves are developing in early spring. The fruit, which is a small drupe, is quite sweet.

Celtis occidentalis 'Delta'
Delta Hackberry

Size	10 m.
Form	High-headed, upright-oval.
Habit	Decurrent.
Canopy	Dense.
Texture	Coarse.
Foliage	Green, 5–15 cm long. Leaves are long and pointed, with an asymmetric base.
Bark	Greyish brown, very rough, warty and hard.
Fruit	Fleshy, dark purple, pea-sized.

This cultivar prefers deep soils with adequate moisture. It was selected from a stand growing near the south shore of Lake Manitoba and has been suggested as a possible replacement for the American elm.

Celtis occidentalis 'Delta'

Clematis

Clematis, Virgin's-Bower

The queen of the climbers! Clematis is much prized for its blossoms and sometimes for its feathery fruit. It grows well in sunny locations where its roots can be kept cool. Gardeners will often grow something like lily-of-the-valley or bergenia at the base of the plant to provide the shade necessary for a cool root-run.

Leaves are usually compound with three leaflets. Flowers are apetalous with four colorful sepals.

Most failures with clematis occur during the first year, due to either a lack of attention, to poor weak plants, or to a combination of both. A little attention and some protection during this very critical period generally pays off. In the north clematis can be expected to kill back to the ground every winter.

The pruning of clematis is important if success is to be achieved. In the prairie provinces two plant types may be encountered: those like C. x *jackmannii*, which bloom only on **new wood** starting in late June, and those like Nellie Moser which bloom on **old wood** in spring and later, about midsummer, on **wood of the current season.** All stems from plants belonging to the first of these two types are cut back hard each spring to assure an abundance of new flowering wood. Blossoming of the second group is fostered by a two step management system. Those stems that have produced the **late blossoms** are cut back hard so that they will produce an abundance of new wood next spring. The remaining flowering stems, those that produced **the early blossoms**, are simply "dead headed" as the blossoms fade. These stems will produce flower buds for next year in the latter part of the growing season so they must be retained. The only real pruning on these is confined to the removal of weak or very old stems.

Since the stems that will be responsible for early blossoming next year will have their flower buds already in place, you may consider it prudent to lay these on the ground in the fall to better protect flower buds from low temperature injury.

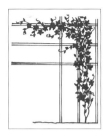

Clematis × *jackmannii*

Jackman Clematis

Size	4 m.
Form	Climber.
Texture	Coarse.
Foliage	Leaves green, compound, the upper ones frequently simple.
Flowers	Deep purple, in threes, 12–15 cm in diameter; sepals fully reflexed.

Plants require a trellis for support; leaf petioles will wrap themselves around the support structure. Sunny exposures are preferred. Cut back heavily in early spring to assure an abundance of flowering wood.

Clematis × jackmannii

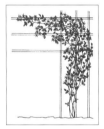

Clematis ligusticifolia
Western White Clematis

Size	4 m.
Form	Climber.
Texture	Medium.
Foliage	Leaves yellowish green, leaflets often three-lobed.
Flowers	Flowers are white, 2 cm in diameter, imperfect/dioecious.

When only female plants are grown, the feathery seed heads will not be produced unless pollen is provided from some other clematis.

Clematis ligusticifolia

Clematis macropetala
Bigpetal Clematis

Size	2 m.
Form	Climber.
Texture	Coarse.
Foliage	Attractive, deeply divided leaves.
Flowers	Violet with conspicuous petaloid staminodes.

This plant was used successfully by the late Dr. Frank Skinner of Dropmore, Manitoba, in the development of hardy, large-flowered, woody climbers.

Clematis macropetala 'White Swan'

Clematis macropetala Hybrids

× **'Rosy O'Grady'** A large pink-flowered climber, this plant blooms on both old and new wood – in early summer on old wood and in late summer on new. Flowers are large and downward-facing. Relay pruning is recommended for this plant.

× **'White Swan'** This is a vine with large, 12.5-cm, white, nodding flowers. It is another clematis that blooms on both old and new wood. Relay pruning is recommended.

Other Hybrids

Large-flowered

× **'Blue Boy'** A climber to 2 m, this cultivar produces an abundance of 60cm flat, outward-facing blue flowers in mid-summer. It blooms on new wood for roughly two months.

× **'Golden Cross'** A strong climber to 6 m, this plant is very floriferous with open faced, yellow flowers produced on new wood. It is late flowering, blooming early from early August until frost.

× **'Nellie Moser'** Very free flowering large-flowered cultivar, pale mauve-pink, each sepal with a carmine central bar; plants are best grown against an east wall or in a shady location to prevent bleaching. Flowers produced in June and again in August and September.

Clematis macropetala

Small-flowered

× **'Grace'** This is a floriferous, small-flowered vine. The flowers are creamy white.

× **'Pamela'** This is a strong-flowering, small-flowered climber. Flowers are white and bloom from mid-summer to frost.

Clematis recta
Ground Clematis

Size	0.75 m.
Form	Sprawling sub-shrub.
Texture	Coarse.
Foliage	Green.
Flowers	Small, creamy white, fragrant flowers freely produced over most of the summer.

This plant will make a neat ball-shaped plant when supported with a wire hoop. Since it blooms on new wood it should be cut back in early spring. The botanical variety *mandshurica* is taller and has larger flowers.

Clematis recta

Clematis tangutica
Golden Virgin's-Bower

Size	3 m.
Form	Climber.
Texture	Coarse.
Foliage	Green, leaves coarsely toothed.
Flowers	Bright yellow, nodding, sepals not reflexed.
Fruit	Single-seeded with a silky feather-like appendage.

This is a strong growing plant of easy culture but not recommended if neatness is valued. Flowers bloom from early summer onwards, and the plant requires a support structure.

Clematis tangutica

Woody Ornamentals

Clematis verticillaris

Clematis verticillaris var. *columbiana*
Western Blue Clematis

Size	3 m.
Form	Climber.
Texture	Coarse.
Foliage	Three leaflets per leaf, coarsely-toothed or entire.
Flowers	Bluish purple, 3–5 cm.

An attractive native climber but one that is seldom seen in cultivation.

Cornus

Dogwood

The hardy species and cultivars of dogwood are, with two exceptions, all shrubs. Some of the species have outstanding winter bark in red, yellow, purple, or bright scarlet. Those species grown for winter bark should have several of the oldest branches removed at ground level each spring to encourage new brightly-colored shoots to emerge. Several cultivars are grown for their variegated foliage. The large-flowered tree types are not hardy in the prairie provinces.

Cornus alba

Cornus alba
Tatarian Dogwood

Size	1 m.
Form	Globular.
Texture	Coarse.
Foliage	Soft green.
Flowers	Yellowish white.
Bark	Blood red.
Fruit	Bluish white.

A hardy species with no special needs. Many good selections are available.

Cultivars of *Cornus alba*

'Argenteo-marginata' (silver-leaved dogwood) A variegated form, white on green with white predominant. It is very showy and combines particularly well with red and purple-foliaged plants. There appears to be another clone of this type sold under the name; however, it is a taller plant and the white variegation tends to be confined to a narrower band along the margins of the leaf. It is not nearly as attractive.

'Aurea' A dense, globe-shaped shrub with yellow-green leaves.

'Gouchaultii' Not as upright as *'Argenteo-marginata'*. Leaves are variegated, yellow and green with the occasional bit of rose-purple. It is more manageable in a sunny spot.

'Kesselringii' (purple-twig dogwood) An upright-spreading plant with bronzy-green foliage and attractive deep purple stems.

'Sibirica' (Siberian coral dogwood) A plant with clean, upright, bright red stems and soft green, deeply-veined foliage. Leaves are oblong with acuminate tips. Fruit clusters are blue-white. It should be combined with other things to exploit the strong winter effect of its bark color.

Cornus alba 'Sibirica'

Cornus alternifolia
Pagoda Dogwood

Size	3 m.
Form	Upright-spreading.
Texture	Coarse.
Foliage	Green, leaves tend to cluster near the tips of branches.
Flowers	White.

This is an attractive large shrub or small tree. A good landscape plant but one not commonly seen in cultivation. It has a distinct horizontal branching habit, hence the common name. It is native to Manitoba and Saskatchewan, and worthy of wider use. As the specific name suggests, the leaves are alternate rather than opposite.

Cornus alternifolia

Cornus canadensis
Bunchberry

Size	0.1 m.
Form	Low, ground-hugging plant.
Texture	Coarse.
Foliage	Attractive light green with deep veins.
Flowers	White.
Fruit	Size of a small pea, bright scarlet, borne in clusters.

This is an attractive groundcover for shady, moist areas in soils containing decayed organic matter. It spreads from slender horizontal rootstalks. It is not aggressive and therefore combines well with other things.

Cornus canadensis

Cornus rugosa

Cornus rugosa
Rugose-Leaved Dogwood

Size	2 m.
Form	Upright.
Texture	Coarse.
Foliage	Green, deeply-veined.
Bark	Purplish.

This plant should be managed so that new wood is encouraged. Young bushy plants look best. The foliage of this plant is its outstanding quality; it is very much like that shown by the best of the tree species but even more attractive.

Cornus sericea 'Isanti'

Cornus sericea 'White Gold'

Cornus sericea
Red-Osier Dogwood

Size	2–3 m.
Form	Mound-like.
Texture	Coarse.
Foliage	Green, but coloring up nicely to red in autumn.
Flowers	Small, white, in clusters, borne intermittently over a long season.
Bark	Dull red, becoming scurfy with age.
Fruit	White, sometimes bluish.

This plant is easy to grow but not satisfactory where space is limiting. It is a good shrub for reclamation purposes. It does best in moist to wet habitats and develops best fall color when grown in full sun. Branches root readily when in contact with moist soil.

Cultivars of *Cornus sericea*

'Flaviramea' A neat plant with many upright golden-barked stems. It has been known to tip-kill in some areas.

'Isanti' A compact selection to 2 m that is well worth using where space is limited.

'Kelseyi' A dwarf type with rather weak stems. Winter effect is poor. It is suitable only in mass plantings or as a hedge plant.

'White Gold' An excellent white-on-green, variegated dogwood. It is even more showy than *C. alba* 'Argenteo-marginata'. Stem color is bright gold.

Coronilla

Crown Vetch

Floriferous, sprawling perennial herbs with flowers borne in axillary peduncled heads or umbels. Flowers are produced freely and over a long season from late spring until late summer. The species described is widely used in many parts of the eastern U.S. as a reclamation plant to prevent erosion.

Coronilla varia
Crown Vetch

Size	0.4 m.
Form	Mound shaped.
Texture	Fine.
Foliage	Leaves are compound, typical of other vetch species.
Flowers	Pink.

This attractive groundcover plant is good for controlling erosion and weeds on roadsides. It is a useful plant in the larger, more natural landscape.

Coronilla varia

Corylus

Hazelnut

The hazelnuts are medium to coarse-textured plants that bear nuts in bristly involucres. Leaves are hairy. Members of the genus are monoecious. The staminate flowers are borne terminally in slender catkins much before the leaves open. The pistillate flowers arise from clusters of buds on short branches. They are very attractive, but are so small their beauty is generally overlooked.

Corylus cornuta
Beaked Hazelnut

Size	2 m.
Form	Upright-spreading.
Texture	Medium.
Foliage	Green.
Flowers	Pistillate flowers are deep red, but inconspicuous.
Fruit	A nut with a husk that becomes beak-shaped at the tip.

While not considered a useful plant for cultivated landscapes, it may have some place in plantings for wildlife.

Corylus cornuta

Woody Ornamentals

Cotoneaster

Cotoneaster

The cotoneasters are a group of medium-textured ornamental shrubs bearing red or dark blue fruit. The fruit is not juicy and the flesh is dry, soft and very much like that of pumpkin or squash. Species vary from low-prostrate forms to upright-spreading types. The genus is moderately susceptible to the diseases fireblight and silver leaf and very subject to attack by the pear slug, the larva of the pear sawfly. This insect appears in mid-summer and rasps the upper epidermis from the leaves, leaving an unsightly mess and preventing the plants from developing the good fall color for which some species are highly valued. Scurfy scale (*Chionaspis furfura*) has also been troublesome on cotoneaster in recent years.

Cotoneaster adpressus 'Praecox'

Cotoneaster adpressus

Cotoneaster adpressus
Creeping Cotoneaster

Size	0.15 - 0.3 m.
Form	Prostrate.
Texture	Fine.
Foliage	Small, oval leaves with wavy margins.
Fruit	Sub-globose, bright red.

This is a useful woody groundcover with clean, glossy leaves. Open, sunny locations are preferred. It is sometimes used on earthen mounds to provide visual articulation to the ground form. When used in mass, plants must be placed close together in order that they may compete with weeds during the establishment period.

Cultivars of *Cotoneaster adpressus*

'Praecox' (Nan Shan Cotoneaster) This plant has a larger, more arching habit than the species, but it is similar in all other respects.

Cotoneaster integerrimus
European Cotoneaster

Size	1 m.
Form	Mound-like.
Texture	Medium.
Foliage	Dark green above, woolly beneath.
Bark	Reddish brown.
Fruit	Bright red.

The European cotoneaster exhibits a sprawling branch structure after the leaves have fallen. It grows well in most environments.

Cotoneaster integerrimus

Cotoneaster lucidus
Hedge Cotoneaster, Peking Cotoneaster

Size	2 m.
Form	Upright-spreading, closed to base.
Texture	Medium.
Foliage	Dark green, shiny above, woolly beneath, excellent bright orange-red fall color.
Fruit	Dark blue to black.

This vigorous, hardy shrub is widely used as a hedge plant. Older hedges tend to become infested with scurfy scale, which must be controlled if the plants are to survive. Cotoneaster hedges generally require renewal every 15 years. For hedges, plant at 30 cm on center.

Cotoneaster lucidus

Cotoneaster racemiflorus 'Soongoricus'
Sungari Rock Spray Cotoneaster

Size	2 m.
Form	Ball-shaped.
Texture	Medium.
Foliage	Grey-green.
Flowers	White, conspicuous.
Fruit	Red.

Arching branches, interesting foliage color and bright red fruit make this a worthwhile ornamental.

Cotoneaster racemiflorus 'Soongoricus'

Woody Ornamentals

Cotoneaster rotundifolius
Roundleaf Cotoneaster

Size	0.3 m.
Form	Low-spreading.
Texture	Medium.
Foliage	Round, green and shiny, turning red in autumn.
Fruit	Bright red.

This plant has an interesting herring bone branching habit, red fruit clusters and good fall color. It does best in sheltered locations where it can receive some sunlight.

Cotoneaster rotundifolius

Cotoneaster submultiflorus
Flowering Cotoneaster

Size	3 m.
Form	Upright-spreading shrub with arching-branches; closed to base.
Texture	Medium.
Foliage	Grey-green.
Flowers	Conspicuous.
Fruit	Bright red, in clusters.

This large attractive shrub masses well. It is very much like the Sungari rockspray cotoneaster. Because of its arching branches, it is a plant that lends itself to being pruned into the form of a small weeping tree. The species is of easy culture.

Cotoneaster submultiflorus

Cotoneaster tomentosus
Brickberry Cotoneaster

Size	2 m.
Form	Upright-spreading.
Texture	Medium.
Foliage	Dull grey-green on upper surface, hairy beneath.
Fruit	Dull brick red.

Fruit color is the major feature of this plant. It is inferior in ornamental value to the other large-sized, red-fruited species.

Cotoneaster tomentosus

Crataegus

Hawthorn

The hawthorns used in the region are small trees or large shrubs. They are useful materials for large, open, park-like sites where they can be very effective in small groups. Leaves are simple, entire or deeply cleft. The fruit is small, but may attain the size of a small crabapple in some species. It is usually dry and flat-tasting with a large seed-to-flesh ratio. Winter buds are lustrous and leathery looking.

Most hawthorns and apples are alternate hosts for cedar/apple rust, a disease which spends part of its life cycle on juniper and cedar. Hawthorns are also subject to attack by the pear slug in late mid-summer each year. Most selections also have strong, sharp thorns.

Crataegus arnoldiana
Arnold Hawthorn

Size	4 m.
Form	Low-headed.
Habit	Decurrent.
Canopy	Dense.
Texture	Medium.
Foliage	Glossy green.
Flowers	White.
Fruit	0.5 cm, scarlet, abundant.

An attractive, small tree with a good branching habit.

Crataegus arnoldiana

Crataegus × *mordenensis*
Morden Hawthorn

Size	3–4 m.
Form	Upright-oval, low-headed.
Habit	Decurrent.
Canopy	Dense.
Texture	Medium.
Foliage	Dark-green, glossy, coarsely-toothed leaves.
Flowers	Double, white, sometimes with a faint tinge of pink.
Fruit	Crimson, produced sparingly.

An attractive flowering tree but not fully hardy in all parts of the region.

Crataegus × *mordenensis* 'Toba'

Cultivars of *Crataegus* × *mordenensis*

'Snowbird' This white-flowered form is hardier than the 'Toba' hawthorn and highly resistant to cedar/apple rust.

'Toba' This is a neat, attractive, small, flowering tree. Although it is of borderline hardiness in the parts of the region where summers are short and winter conditions more severe, it is deserving of much wider use. It is highly resistant to cedar/apple rust.

Crataegus × *mordenensis* 'Toba'

Crataegus succulenta

Crataegus succulenta
Fleshy Hawthorn

Size	5 m.
Form	Low-headed.
Habit	Decurrent.
Canopy	Dense.
Texture	Coarse.
Foliage	Glossy green.
Flowers	White, showy in large flat-topped corymbs.
Fruit	Bright red, sticky. Of the hardy species this one has the largest fruits.

This is a good ornamental, but it is highly susceptibile to cedar/apple rust.

Cytisus

Broom

These small, showy, spring-blooming shrubs produce an abundance of pea-like flowers on low, prostrate, upright or mound-forming plants. Flowers are generally yellow; however, one of the species described has purple blossoms. The brooms do best in open sunny locations. They thrive in the poorest of soils.

Cytisus pilosa 'Vancouver Gold'
Rock-Garden Broom

Size	0.1 m.
Form	Low-spreading plant.
Texture	Fine.
Foliage	Green.
Flowers	Small, bright yellow, produced in great abundance.

This is a very attractive, small plant which could be very useful as a flowering groundcover in protected sunny locations.

Cytisus pilosa 'Vancouver Gold'

Cytisus purpureus
Purple Broom

Size	0.4 m.
Form	Low-spreading plant.
Texture	Fine.
Foliage	Green.
Flowers	Lilac-purple.

This plant is suited to open sunny locations. Its blossoms, which are produced in early June, are very attractive.

Cytisus purpureus 'Albus'

Cytisus ratisbonensis
Golden Broom

Size	1 m.
Form	Mound-like.
Texture	Fine.
Foliage	Light green, palmately compound.
Flowers	Bright sulfur-yellow in loose arching racemes.

This is a very floriferous plant that is most attractive at time of bloom. Its habit of growth is somewhat delicate with graceful arching branches. It is susceptible to a leaf-spot disease which can result in moderate leaf drop.

Cytisus ratisbonensis

Woody Ornamentals

Daphne

Daphne, Garlandflower

The daphnes are small very attractive flowering shrubs with fragrant blossoms that are produced in axillary or terminal clusters. Both deciduous and broad-leaved evergreen species have proven hardy on the prairies. Flowers are of small to medium-size; color is variable with lilac, pink and yellow forms existing. Daphnes do best on well drained soils. It has been said that they have a preference for calcareous soils but this is not confirmed. The leaves and fruits of all daphnes are poisonous.

Daphne cneorum

Daphne cneorum
Rose Daphne

Size	Less than 0.3 m.
Form	Low-spreading, broad-leaved evergreen.
Texture	Fine.
Foliage	Leaves linear, light green, in whorls.
Flowers	Small, in many flowered clusters, deep rose-pink. Main flowering period is late spring.

Rose daphne is very attractive as a groundcover or in a small grouping. Because it is a broad-leaved evergreen it requires lasting snow cover to protect the foliage from the effect of early spring sunshine. It does best in well-drained soils where moisture can be provided.

Daphne mezereum

Daphne mezereum
February Daphne

Size	0.9 m.
Form	Oval to ball shaped.
Texture	Medium.
Foliage	Medium-green deciduous foliage in whorls about the branches.
Flowers	Small lavender flowers that arise directly from the upright stems. Flowers are produced before the leaves.
Fruit	Soft, juicy, scarlet drupes (8 mm) which are held tightly along the stems from top to bottom. The fruit is poisonous.

This is one of the more attractive small shrubs for landscape purposes. It is the earliest of our woody plants to blossom in spring.

Duchesnea

Duchesnea

These ground hugging perennials are often mistaken for the strawberry because of the similarity that exists in foliage, form, and fruit of the two genera. While the fruit of *Duchesnea* resembles that of the wild strawberry it is tasteless.

The flowers of *Duchesnea* are yellow rather than white, and calyx segments are prominent, bract-like structures. *Duchesnea* makes a good groundcover for open, sunny locations.

Duchesnea indica
Yellow-flowered Strawberry

Size	0.1 m.
Form	Ground-covering herbaceous perennial that spreads by above-ground runners (stolons).
Texture	Medium.
Foliage	Dark green, trifoliate, like strawberry.
Flowers	Yellow, the calyxes are subtended by a whorl of bracts that are larger than the calyx segments and alternate with them.
Fruit	Red, like strawberry, but insipid to taste.

An excellent groundcover that establishes itself quickly from leafy runners.

Duchesnea indica

Elaeagnus

Oleaster

This genus contains deciduous trees and shrubs that are well suited to dry land and mildly saline soil conditions. The silvery leaf color of the plants is their most outstanding characteristic; the color is due to the presence of silvery scales that cover the leaves, fruit, and new stem growth of most species. It has been reported that members of the genus are not likely to be hardy in the northern parts of the region.

The flowers of *Elaeagnus* have a heavy fragrance. The fruit is a large-seeded drupe with dry flesh.

Woody Ornamentals

Eleagnus angustifolia

Elaeagnus angustifolia
Russian-Olive

Size	4–5 m.
Form	Low-headed, upright-oval.
Habit	Decurrent.
Canopy	Open.
Texture	Fine.
Foliage	Silvery
Flowers	Yellow, very fragrant.
Fruit	Dry, silvery, with a single oblong seed.

Because of its foliage color, the Russian-olive contrasts nicely with conifers. The foliage also tends to be retained into the winter. The root system of this tree may be subject to winter injury some years.

Eleagnus commutata

Elaeagnus commutata
Wolf-Willow, Silverberry

Size	2–3 m.
Form	Upright, thicket-forming native plant.
Texture	Medium.
Foliage	Strap-shaped, silvery.
Flowers	Yellow, very sweet scented.
Fruit	Globular, dry, silvery, one seeded.

This plant is commonly found growing on gravelly river-bench soils. The form is not attractive and the plant is difficult to use in combination with other plants but still it can be used in the landscape should it exist on the site. It has been used as a hedge plant.

Euonymus

Burningbush, Spindletree, Strawberrybush

Deciduous or broad-leaved evergreen shrubs. They are popular because of their autumn coloration which will vary with most species from bright pink to bright red. Some species may fail to achieve their full autumn color in this region because leaf chlorophyll is sometimes slow to break down. Fruit is small and somewhat unusual. It is a capsule, usually lobed and sometimes winged. When ripe the capsule splits open revealing a bright orange inner section, the aril.

Euonymus alatus
Winged Burningbush

Size	1.5 m.
Form	Upright-spreading, leggy.
Texture	Medium.
Foliage	Green.
Bark	Branches and twigs have strange corky ridges which protrude from the bark, giving the branches a squarish cross-section.
Fruit	A small, but attractive capsule which opens when ripe to expose the colorful aril that covers the seed.

The early autumn coloration of this plant is a striking deep pink. Good fall color may be achieved even in semi-shade.

Euonymus alatus

Cultivars of *Euonymus alatus*

'Compactus' This can be an outstanding dwarf shrub because of its neatness and its exceptional deep red fall color. It has been noted that intense sunlight results in the production of "sun-leaves" which are thicker than normal and do not color up to the same degree as the normal or "shade-leaves." Another reaction to strong light is the tendency for plants to produce several thick, straight non-branching stems. To avoid such problems, plants should be grown in diffused light.

Euonymus alatus 'Compactus'

Euonymus europaeus
Spindletree

Size	To 2 m.
Form	Upright-spreading.
Texture	Coarse.
Foliage	Green.
Bark	Green.
Fruit	Bright pink capsules produced in abundance.

This plant prefers full sun. Deep pink autumn foliage and showy fruit are its outstanding characteristics.

Euonymus europaeus

Cultivars of *Euonymus europaeus*

'Aldenhamensis' This is a neat ball-shaped shrub to 1 m, but it has little else to offer other than size and form. Its leaves do not color up like other members of the genus.

Euonymus fortunei 'Vegetus'

Euonymus fortunei 'Vegetus'
Wintercreeper

Size	0.6 m.
Form	Climbing plant.
Texture	Medium.
Foliage	Glossy, dark green leaves that are retained over winter when the protection of snow cover is provided.
Flowers	Greenish white, in clusters.
Fruit	Pink, splitting open to expose an orange aril.

This little climber has survived in Edmonton against the west wall of a building for many years and on the basis of this experience it should be tested more widely. It requires a moist soil with a moderate amount of organic matter.

Euonymus maackii
Maack's Spindletree

Size	3 m.
Form	Upright-spreading.
Texture	Coarse.
Foliage	Green.
Flowers	Yellowish.
Fruit	Pink to red
Bark	Dappled-grey.

This is a good large shrub or small tree. It lacks the bright red fall color of the better spindle-trees but does color up to a bright orange.

Euonymus nanus 'Turkestanicus'

Euonymus nanus
Dwarf Narrow-Leaved Burningbush

Size	0.6 m.
Form	Upright, but tending to sprawl.
Texture	Fine.
Foliage	Semi-evergreen linear leaves, dark green turning pink in autumn.
Fruit	Deep pink in clusters, showy.

This plant does not have widespread use because of its sprawling habit, but it could be useful for massing on slopes where grass is not a practical solution. When used as a groundcover, the recommended planting distance is 15 cm on center.

Cultivars of *Euonymus nanus*

'Turkestanicus' Similar to the species but more upright in habit and much taller (to 1.5 m) and denser. Globular specimens have also been noted. Colorful fruits are very showy in late summer.

Euonymus nanus 'Turkestanicus'

Euonymus obovatus
Running Strawberrybush

Size	Less than 0.3 m.
Form	Procumbent, spreading.
Texture	Medium.
Foliage	Light green.
Fruit	Crimson with scarlet aril; warty.

This vigorous, hardy, woody groundcover has stiff, straight stems that arise from the crown at odd angles. In more temperate climates foliage turns red in the autumn. Autumn color changes are not common in this region. The plant does well in full-sun or semi-shade.

Euonymus obovatus

Euonymus verrucosus
Warty-Bark Burningbush

Size	1 m.
Form	Neat, globular.
Texture	Medium.
Foliage	Green, with outstanding fall color.
Bark	Stems, particularly the new growth, are covered with dark wart-like glands.
Fruit	A brown capsule splitting to reveal an orange aril.

This plant is fully hardy and does best in full sun. It provides an outstanding display of deep pink foliage color in early fall.

Euonymus verrucosus

Woody Ornamentals

Festuca

Fescue

A fine-textured grass with potential value as a groundcover with a difference. The species described is a hummock-forming plant that makes a good ornamental because of its color and form. It is useful in both sun and shade.

Festuca ovina 'Glauca'

Festuca ovina 'Glauca'
Blue Sheep's Fescue

Size	0.3 m.
Form	Low hummock or mop-head.
Texture	Fine.
Foliage	Silvery blue.

This plant can be used to provide low hummocks of grassy vegetation. It is usually planted on 20-cm centers. It may be more acceptable if treated as an annual plant or cut back severely in the fall since it bleaches badly during the winter and looks unsightly in early spring before growth resumes.

Forsythia

Forsythia, Goldenbells

These hardy spring-flowering shrubs are rare in the region because flower buds are frequently subject to winter-kill when snow cover is inadequate. Cultivars of the one species described tend to produce hardier flower buds than those of species commonly seen on the west coast and in southern Ontario. In spite of the risk, *Forsythia* is well worth trying in sheltered locations where snow cover can be assured. The flowers are a golden-yellow and appear before the leaves.

Forsythia ovata

Forsythia ovata
Korean Goldenbells

Size	2 m.
Form	Upright, leggy.
Texture	Medium.
Foliage	Green.
Flowers	Golden-yellow, very early.

The value of this small to medium-sized shrub is exclusively due to the spring warmth conveyed by its blossoms which appear while most other woody plants are still dormant. Flower buds that are exposed above the snow line are injured most years.

Cultivars of *Forsythia ovata*

× **'Northern Gold'** A large-sized shrub with flower buds that are hardier than those of the species.

× **'Ottawa'** A reliable flowering plant with a compact habit of growth.

Forsythia ovata 'Northern Gold'

Fraxinus
Ash

Ash trees are tall, deciduous, and long-lived with handsome foliage. The leaves are compound and opposite-pinnate (like a feather). The bark is smooth when young but roughens with age. Flowers are inconspicuous, imperfect/dioecious; fruit is dry and single-seeded with an elongated wing. In most cases the unattractive fruit is retained over winter; hence male clones are preferred. Ash species tend to be late starters in the spring and are quick to lose their leaves in the fall.

Fraxinus americana
White Ash

Size	12 m.
Form	Upright-oval.
Habit	Decurrent.
Canopy	Dense.
Texture	Coarse.
Foliage	Green, but turning deep purple in autumn.

Moist sites are preferred. This is a neat, fairly narrow tree during its early years. As it gets older it becomes a good shade tree with a stature much like that of green ash.

Fraxinus americana

Cultivars of *Fraxinus americana*

'Autumn Blaze' A female clone that appears well adapted to most regional conditions.

Woody Ornamentals

Fraxinus mandshurica

Fraxinus mandshurica
Manchurian Ash

Size	8 m.
Form	Compact, upright-oval.
Habit	Decurrent.
Canopy	Dense, closed.
Texture	Coarse.
Foliage	Light green.
Bark	Grey, furrowed on older trees.

The Manchurian ash is fully hardy on sites where good soil conditions exist. Because of its compact form, it can be considered a useful street tree.

Fraxinus nigra

Fraxinus nigra 'Fallgold'

Fraxinus nigra
Black Ash

Size	10 m.
Form	High-headed, narrowly-pyramidal.
Habit	Excurrent.
Canopy	Dense.
Texture	Coarse.
Foliage	Green, turning bright yellow in early fall.

This tree is native to eastern Manitoba, where it can often be seen growing in standing water. In spite of its tolerance of wet conditions, it will adapt to drier locations but would likely suffer in the drier parts of the region. Because of its excurrent habit, the black ash tends to make a good, easily managed street tree.

Cultivars of *Fraxinus nigra*

'Fallgold' A male clone with a narrow pyramidal head and longer lasting fall color. A good street tree because of its uniformity.

Fraxinus pensylvanica var. *subintegerrima*
Green Ash

Size	12 m.
Form	High-headed, upright-oval.
Habit	Decurrent.
Canopy	Dense.
Texture	Coarse.
Foliage	Green. Fall color bright yellow.
Bark	Deeply furrowed on older trees.

Fraxinus pensylvanica var. *subintegerrima*

This tree is widely used over the whole region as a tall shade and/or street tree. Crowns of older specimens tend to produce a surplus of wood and thus require a lot of maintenance when they get older. Foliage is subject to attack by the lygas bug (*Lygus lineolaris*) in all parts of the region. In Manitoba and Saskatchewan, canker worms remain a serious problem. Heavy infestations of ash flower gall mite (*Aceria fraxiniflora*) can also create an unsightly appearance on the branches in early spring.

Cultivars of *Fraxinus pensylvanica* var. *subintegerrima*

'Patmore' A male clone that is known to leaf out earlier and retain its leaves longer than other members of the species.

Fraxinus pensylvanica var. *subintegerrima*

Galium
Asperula

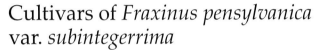

The species *Galium odoratum* is an outstanding groundcover for shady places. It has whorled leaves which are covered with sprays of tiny white fragrant blossoms in late spring. Plants establish themselves quickly and are easily propagated from small divisions.

Galium odoratum
Sweet Woodruff

Size	Less than 0.3 m.
Form	Upright, spreading.
Texture	Medium.
Foliage	Whorls of semi broad-leaved evergreen leaves.
Flowers	White scented flowers, borne above the leaves in early June.

This plant is a vigorous groundcover that will establish itself quickly without difficulty. The leafy habit is very attractive and the scented flowers are highly valued. The plant also used as a flavoring herb.

Galium odoratum

Woody Ornamentals

Genista

Woadwaxen

Small, straight-stemmed sub-shrubs or low spreading plants that are grown for their bright yellow, pea-like flowers. The genus is related to *Cytisus* and as is the case with *Cytisus* the *Genistas* do very well in well-drained, sandy, or infertile soils, in full sun. Plants are difficult to transplant, so those that are established cannot be moved easily.

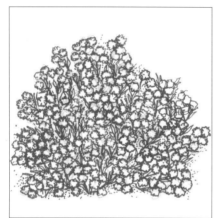

Genista lydia

Genista lydia
Lydia Woadwaxen

Size	0.6 m.
Form	Mound-like.
Texture	Fine.
Foliage	Light green.
Flowers	Golden-yellow.

This is an outstanding dwarf, flowering plant. The slender pendulous branches are covered with flowers in late spring.

Genista tinctoria 'Rossica'

Genista tinctoria
Dyer's Greenweed

Size	0.6 m.
Form	Upright-spreading.
Texture	Fine.
Foliage	Light green.
Flowers	Bright yellow.

This plant flowers throughout the summer.

Cultivars of *Genista tinctoria*

'Rossica' Similar to the species but a more vigorous plant.

Ginkgo

Maidenhair Tree

A genus of very ancient trees of which only a single species survives. In the prairie region this tree is very rare though quite common in more moderate climates. Most leaves are borne in short spurs. The flowers are imperfect/dioecious. The staminate flowers are in catkins, the pistillate ones on long pedicels. The fruit has a pulpy outer layer that is high in oils and smells strongly, like rancid butter, when ripe. For this reason female trees should be avoided.

Ginkgo biloba
Maidenhair Tree

Size	5 m.
Form	Upright-spreading.
Habit	Decurrent.
Canopy	Open.
Texture	Medium.
Foliage	Unusual fan-shaped leaf with veins following an open, forking system without any vein fusions. Autumn color is bright yellow.
Flowers	Imperfect/dioecious.

This tree is very slow-growing and very rare. A specimen in Edmonton has not grown to more than 3 m in 10 years though there has been no evidence of winter-kill. Botanists have considered *Ginkgo* a living fossil since it is a representative of an ancient order of conifer-like trees that existed in the Mesozoic era. Not recommended, but worthy of trial.

Ginkgo biloba

Halimodendron

Salt-tree

The one species belonging to this genus is a plant that is native to the dry, saline parts of Siberia. Like many of our own halophytes, this plant has silvery leaves and sharp spines. Because of its tolerance of dry, saline soils, it is a good plant for Prairie Canada. Like *Caragana*, *Halimodendron* is a member of the pea family, *Fabaceae*.

Woody Ornamentals

Halimodendron halodendron
Salt-tree

Size	2 m.
Form	Upright-spreading, leggy.
Texture	Fine.
Foliage	Grey-green.
Flowers	Pale lilac, attractive.
Fruit	A 2.5-cm inflated pod.

This tall shrub is suited to saline soils. It is a handsome ornamental when in bloom. It is sometimes grafted on common caragana.

Halimodendron halodendron

Hedera

Ivy

Broad-leaved evergreen groundcover plants for full or partial shade. Members of the genus are more commonly thought of as vigorous climbers but ones that would have great difficulty surviving in a north temperate winter. In the prairie provinces a favorable microclimate is essential to their survival as groundcovers. The single cultivar described is the only one that can be recommended at this time.

Hedera helix 'Baltica'
Baltic Ivy

Size	0.15 m.
Form	Creeping plants with a tendency to climb.
Texture	Medium.
Foliage	Leaves lobed, thick, dark glossy green.

This groundcover is very attractive and fresh-looking, but must be grown in a sheltered area where it can be assured snowcover and winter soil temperatures that do not go below freezing. It survives well close to house foundations.

Hedera helix 'Baltica'

Hippophae

Sea-buckthorn, Russian Sandthorn

Large, upright shrubs with distinctive, narrow, linear leaves that are silvery green in color and noticeably fine in texture. As with *Elaeagnus* the leaves of *Hippophae* are alternate and covered on both sides with silvery scales. The flowers are imperfect/dioecious and, while not considered showy, the pistillate ones give rise to clusters of bright orange fruit that cover shoots produced the previous year. Plants have some tendency to sucker when their roots are disturbed. Stems and lateral shoots terminate in sharp thorns.

Hippophae rhamnoides
Common Sea-buckthorn

Size	4 m.
Form	Upright-spreading, leggy.
Texture	Fine.
Foliage	Grey.
Flowers	Imperfect, inconspicuous.
Fruit	1.5–2 cm, single seeded, juicy, yellow to orange and elongate. These are held tightly to the fruiting shoots, effectively covering them.

This is a rugged yet very attractive large shrub with excellent color and texture. Both male and female plants are necessary for fruit production. Fruit persists all winter and can be a very showy element in the winter garden. Fruiting branches are also popular for use in winter bouquets.

Hippophae rhamnoides

Hydrangea

Hydrangea

Medium-sized, coarse-textured shrubs, hydrangeas are grown for their large, showy inflorescences that are borne terminally. Flowering begins in mid-summer and flowers are retained to the end of the growing season. The inflorescence of hydrangea is either a terminal corymb or panicle and consists of both fertile and sterile florets. Moist, semi-shaded habitats are preferred by the two species grown in the region.

Hydrangea arborescens 'Grandiflora'

Hydrangea arborescens 'Grandiflora'
Snow Hills Hydrangea

Size	0.75 m.
Form	Globose, closed to base. Stems stiff, straight and non-branching.
Texture	Coarse.
Foliage	Green.
Flowers	Fertile florets are small and insignificant, but the large, white, somewhat flattened clusters of sterile ones are very showy.

This plant does well in moist soils in shady habitats. It can be very effective when massed. Because the plants have no winter value they are generally cut back to 15 cm in the fall. This has no detrimental effect on the following year's show of blossoms since flower buds are produced only on new wood.

Cultivars of *H. arborescens* 'Grandiflora'

Hydrangea arborescens 'Annabelle'

'Annabelle' A shrub valued for its annual display of large, white, long-lasting flower heads. In this cultivar the flowers are all sterile and are produced on large corymbs up to 25 cm in diameter.

Hydrangea paniculata 'Grandiflora'

Hydrangea paniculata 'Grandiflora'
Pee Gee Hydrangea

Size	1 m.
Form	Upright-spreading, very tidy plant with an attractive and easily managed growth habit.
Texture	Coarse.
Foliage	Leaves smaller and more attractive than those of the snow hills hydrangea.
Flowers	Large pyramidal panicles of pinkish sterile flowers.

This plant is a good flowering shrub for semi-shade. It does not have the same weak winter habit displayed by the snow hills hydrangea and consequently is not handled the same way. Branches are commonly shortened on an annual basis to reduce the number of buds from which new flowering shoots will emerge. This form of management will result in fewer but larger flower clusters. The cultivar 'Praecox' is also sold in the region. It is similar to 'Grandiflora' but has smaller flower clusters that are a mixture of both sterile and fertile florets.

Juglans

Butternut, Walnut

Members of the genus are large-sized trees with large, showy, pinnately compound leaves; flowers are imperfect/monoecious. Male flowers are borne in non-branched catkins. The fruit is a nut. These trees are rare in the region, but those that have been encountered appear to be well adapted to the climate of north central parts. All members of the genus do not transplant well after they reach a certain size; young whips or seedlings can be established more easily than large transplants.

Juglans cinerea
Butternut

Size	10 m.
Form	Upright-spreading.
Habit	Decurrent.
Canopy	Open.
Texture	Medium.
Foliage	Light green, very large compound leaves.
Bark	Grey, furrowed.
Fruit	Ovoid-oblong, husk coated with rusty sticky hairs.

This tree appears to be well-suited to conditions in the north-central part of the region.

Juglans cinerea

Juglans nigra
Black Walnut

Size	10 m.
Form	Upright-oval.
Habit	Decurrent.
Canopy	Open.
Texture	Medium.
Foliage	Green.
Fruit	Globular with a thick husk.

Valued as a large shade tree with attractive foliage.

Juglans nigra

Juniperus

Juniper

Junipers are coniferous shrubs or trees with needles opposite or in threes. Needles are either dull-pointed and scale-like or awl-shaped, and are pungent when crushed. Bark is thin and shreddy. The fruit is a small, blue, berry-like cone with a small bract at the base.

The inability of some junipers to tolerate the early spring sun without injury can limit their use against structures with southern exposures. Nevertheless the junipers still prefer sunny environments.

Juniperus communis

Juniperus communis
Common Ground Juniper

Size	0.6 m.
Form	Semi-prostrate. Branches tend to arch downwards at the tips.
Texture	Fine.
Foliage	Bright green with a silver band on the upper surface of each awl-shaped needle. Needles arranged in 3's.

This is a native plant common to the mountains and foothills of Alberta. The habit of this plant is looser than that of the domesticated junipers. Its foliage becomes a rather sombre purplish color when the cold weather arrives.

Cultivars of *Juniperus communis*

'Depressa Aurea' A low-growing plant with golden-yellow foliage. It is one of the better golden forms of any juniper regardless of species.

'Repanda' This low plant has semi-prostrate stems and dark green, densely-packed foliage, which has some resemblance to whipcord.

Juniperus communis 'Depressa Aurea'

Juniperus horizontalis
Horizontal Juniper

Size	0.15–0.3 m.
Form	Decumbent.
Texture	Fine.
Foliage	Green to silvery blue.

This native species is highly variable and found in the open over a wide area. A large number of selections have been named and are readily available from all nurseries.

Cultivars of *Juniperus horizontalis*

'Blue Chip' An outstanding cultivar with branches radiating from the center of the plant. Bright blue color is retained throughout the year.

'Blue Prince' Plant with a neat prostrate growth habit and a powder-blue foliage color. As a groundcover it can be used to provide good color and textural contrast with the foliage of other plants.

'Douglasii' (Waukegan juniper) Main branches are horizontal, but the branchlets ascend at a steep angle. Foliage has a rich grey green appearance which changes to a soft greyish mauve in autumn.

'Dunvegan Blue' A low, ground-hugging form with exceptionally good silvery blue foliage. It is by far the best of its type.

'Hughes' A good, low-growing, silvery blue plant, but not as ground-hugging as 'Dunvegan Blue'.

'Plumosa' (Andorra juniper) One of the best of the low, mound-forming types. Branches radiate from the center of the plant, giving it a circular configuration. Branches rise at an angle of about 45 degrees. It tends to have a soft, mauve color which intensifies during the latter part of the growing season.

'Plumosa Compacta' A denser selection of Andorra juniper but one that does not stay full at the center. Winter color brownish purple.

'Prince of Wales' A low, 0.1–0.25 m, bright green, ground-hugging plant. Lateral growth is very short; terminal growth is longer. This is an excellent low groundcover.

'Turquoise Spreader' A mat-forming groundcover with attractive and distinctive blue-green foliage, with bright green under-foliage.

'Wapiti' A low juniper to 0.3 m with sharply ascending laterals. The color is green, turning dull purple in autumn.

'Wiltonii' (Blue Rug) This plant has long terminal growth and very short lateral growth. It has a tendency to expose its stems at the crown, giving older plants a rugged, weathered appearance. Its color is glaucous blue. This plant may have some value for the "bonsai" hobbyist.

Juniperus horizontalis 'Blue Chip'

Juniperus horizontalis 'Blue Prince'

Juniperus horizontalis 'Dunvegan Blue'

Juniperus horizontalis 'Hughes'

Juniperus × media 'Pfitzerana Aurea'

Juniperus × media 'Pfitzerana'
Pfitzer Juniper

Size	0.75 m.
Form	Semi-prostrate; appearance soft, feathery.
Texture	Fine.
Foliage	Dark green.

Form and color are attractive. Plants are hardy and often tall enough to be seen above the snow line.

Cultivars of *J. × media* 'Pfitzerana'

'Pfitzerana Armstrongii' A more compact form of the Pfitzerana type. It is highly recommended.

'Pfitzerana Aurea' (golden Pfitzer juniper) Form is typical of the species but terminal growth is a clear, soft yellow. This cultivar is not as hardy as the species; it does much better in parts of the region where winters are mild and summers less dry. Dieback is regularly seen on exposed portions.

'Pfitzerana Compacta' (compact Pfitzer juniper) A smaller plant with very prickly foliage. It is not as hardy as the species and not as widely recommended as some of the other cultivars.

'Pfitzerana Glauca' A silvery-blue form with very prickly foliage.

Juniperus × media 'Pfitzerana Glauca'

'Pfitzerana Mint Julep' A compact form of the species with good, dark green color but lacks hardiness in northern areas.

Juniperus procumbens 'Nana'

Juniperus procumbens 'Nana'
Jap-garden Juniper

Size	0.1 m.
Form	Procumbent, spreading shrub or groundcover with exceptionally good texture and habit.
Texture	Fine.
Foliage	Light green.

This dwarf conifer is quite different from the usual ground-hugging juniper in that it will produce a number of elongating stems which extend to some distance over the ground like the tentacles of some sea creature. Because of its stature, it is best suited to use as a specimen in a rock garden situation.

Juniperus sabina
Savin Juniper

Size	1 m.
Form	Upright with gently arching tips.
Texture	Fine.
Foliage	Dark green, needles chiefly scale-like.

The old savin juniper with its arching branches has outgrown its usefulness but there are many cultivars available, all of which are vastly different and superior to the type.

Cultivars of *Juniperus sabina*

'Arcadia' A small, 0.3–0.6 m, neat, bright-green plant. It is not as upright as the species and its stems arch more gracefully.

'Broadmoor' A low, spreading form with a strong growth habit. Foliage is dull blue-green; plant form is similar to 'Calgary Carpet'.

'Blue Danube' 0.6-0.75 m, a broad, spreading plant of good dark bluish-green color.

'Calgary Carpet' 0.3 m An outstanding introduction. This plant has a pronounced horizontal branching habit which is quite unique for a savin juniper. Its color is a fresh, light green.

'Hoar Frost' A very interesting small plant. The cultivar has an unusual form of growth for a savin cultivar in that it is very tuft-like. In addition, the tips of all new growth are light yellow.

'Skandia' A low shrub (less than 0.3 m) with a slight greyish-green cast to the foliage. This plant is a sister to 'Arcadia' but is much smaller with a more compact habit. It is a good groundcover.

'Tamariscifolia' A tight, rugged, mound-shaped plant to 0.6 m. The strong but stiffly arching branches bear short, upright secondary branches.

Juniperus sabina

Juniperus sabina 'Arcadia' (left)
Juniperus sabina 'Skandia' (right)

Juniperus sabina 'Tamariscifolia'

Juniperus scopulorum 'Blue Heaven'

Juniperus scopulorum
Rocky Mountain Juniper

Size	5 m.
Form	Broad to narrow pyramidal.
Texture	Fine.
Foliage	Green to grey-green.

Rocky Mountain juniper does best on well-drained soils. Young plants have been known to tip-burn in the spring, however once established, they are generally quite sunworthy. Like other juniper species, this one is quite variable and many good selections have been made on the basis of form and color.

Juniperus scopulorum 'Grey Gleam'

Cultivars of *Juniperus scopulorum*

'Blue Heaven' Compact, narrow pyramid with silvery-blue foliage.

'Grey Gleam' A narrow, compact pyramid with good silvery color.

'Grizzly Bear' A broad pyramid, dense with a very neat habit of growth that does not require shearing.

'McFarland' A very narrow columnar form of the species with grey-green foliage. A good substitute for 'Skyrocket'.

'Medora' Compact, narrow pyramid, almost columnar. Its foliage is bluish.

'Moffettii' Broadly pyramidal form with good silvery-blue color.

'Wichita Blue' A broad pyramid with bright blue foliage. It requires shearing to keep it neat.

'Winter Blue' This cultivar does not resemble the typical *J. scopulorum* in form. It is a loose, somewhat sprawling plant to 60–80 cm, with bright-blue foliage.

Larix

Larch, Tamarack

This genus consists of tall pyramidal conifers that replace their needles each year. Needles are produced in clusters on small peg-like spurs in the older growth; on new shoots, needles are borne singly and are spirally arranged.

The larch is particularly beautiful in spring and fall. In spring the needles are a soft, light green and in the fall just prior to leaf drop, the foliage is a clear, golden-yellow. Fall color is striking and it can be accentuated when the trees are grown in association with other conifers like spruce and fir. Because they shed their needles, larches have found greater acceptance as trees in the larger landscape where their winter "nakedness" is more likely to be overlooked.

Larix laricina
Tamarack

Size	12–15 m.
Form	Narrow-pyramid.
Foliage	Light green, turning bright golden-yellow in the fall.
Bark	Smooth grey on young trees, becoming reddish-brown and scaly.
Seed Cones	Small, 1 cm long, egg-shaped, borne erect on short spurs.

The spring and fall color of foliage is the main landscape feature. The tamarack is a swamp tree but it will transplant to dry land sites and grow without difficulty in the more northerly parts of the provinces.

Larix laricina

Larix sibirica
Siberian Larch

Size	30 m.
Form	Broad-pyramid with lower branches descending slowly from the trunk, then arching gracefully upwards toward the tips.
Foliage	Needles 3 cm long, flat, soft. Spring color is a very soft light green; fall color bright yellow.
Bark	Yellow on young trees, becoming light brown with age.
Seed Cones	Egg-shaped, 3 cm long, borne erect on short spurs.

This tree appears to be well adapted to dry land sites. Its habit of growth plus its very attractive spring and fall color make it a good landscape tree. It is fully hardy.

Larix sibirica

Lonicera

Honeysuckle

Honeysuckles are vigorous large or small shrubs and vines. Flowers, which are borne in pairs are produced freely and are quite conspicuous. The fruit is a berry. In most cases berries are borne in pairs. The ornamental fruit is palatable only on *L. caerulea edulis*.

In recent years, two hybrids that display a more or less juvenile growth habit have become available and both have great value as hedge plants. Because of their habit of growth they require shearing much less frequently than hedge-plants that have the mature habit.

In the western provinces, over the last few years, some honeysuckles, particularly cultivars of *L. tatarica* have been attacked by the honeysuckle aphid *(Hydaphis tataricae)* that has had a devastating effect on the appearance of susceptible species and cultivars. The main symptom of the infestation is the fasciated appearance of all blossoms. The insect, an Asiatic native, first appeared in eastern Canada almost twenty years ago but is now firmly established in western Canada. Buyers are strongly advised to check on the susceptibility of each species or cultivar to this pest before making a purchase.

Woody Ornamentals

Lonicera × *brownii* 'Dropmore Scarlet Trumpet'

Dropmore Scarlet Trumpet Honeysuckle

Size	3 m.
Form	Climbing vine.
Texture	Coarse.
Foliage	Lustrous green.
Flowers	Bright orange to red, trumpet-shaped.

One of the best flowering vines for the western provinces and is not subject to injury from the honeysuckle aphid.

Lonicera × *brownii* 'Dropmore Scarlet Trumpet'

Lonicera caerulea var. *edulis*
Sweetberry Honeysuckle

Size	1.5 m
Form	Medium to large mound with stiff arching, purplish branches.
Foliage	Green, oblong.
Flowers	Yellowish white.
Fruit	Oblong, blue, glaucous, taste strong, bitter.

Selections have been made on the basis of fruit quality and the preserved product from these has been found to be quite tasty.

Lonicera caerulea var. *edulis*

Cultivars of *Lonicera caerulea* var. *edulus*

'Lac la Nonne' A large fruit-plant with blueberry-sized fruit of good quality.

Lonicera korolkowii 'Zabelii'
Zabel's Honeysuckle

Size	2 m.
Form	Upright-spreading; not as rugged looking as *L. tatarica*.
Texture	Medium to coarse.
Foliage	Green with a reddish cast.
Flowers	Deep rosy-pink, very showy.

There is some uncertainty as to whether the plant described and sold as Zabel's honeysuckle is correctly named. Nevertheless the plant referred to has been the outstanding one of this type. It is unfortunate that it, like most other honeysuckles, is very susceptible to aphid infestation.

Lonicera korolkowii 'Zabelii'

Lonicera maximowiczii var. *sachalinensis*
Sakhalin Honeysuckle

Size	1.5 m.
Form	Mound-like, closed to base.
Texture	Medium.
Foliage	Green; spring color is soft, lustrous and somewhat copper-colored.
Flowers	Deep purple and showy.

This plant is tolerant of deep shade as well as full sun. The size, form, and color are the main physical attributes of the plant.

Lonicera maximowiczii var. *sachalinensis*

Lonicera spinosa var. *Albertii*
Albert Thorn Honeysuckle

Size	0.6 m.
Form	Sprawling plant, somewhat mound-like with arching branches.
Texture	Fine.
Foliage	Narrow, blue-green.
Flowers	Violet, fragrant.
Fruit	Pale blue or white.

This plant is suited to sunny locations and will not survive winters if grown on wet sites.

Lonicera spinosa var. *Albertii*

Lonicera tatarica
Tartarian Honeysuckle

Size	2–3 m.
Form	Upright-spreading, leggy.
Texture	Medium.
Foliage	Green.
Flowers	Color variable. Cultivars with white, pink, two-toned and deep red flowers are carried by nurseries.
Fruit	Orange or red.

This tall rugged shrub has graced the man-made landscapes of the prairie provinces from the time of early European settlement. The great billowing masses of flowers produced in early June are very striking but the plant has little to commend it when the blossoming period is finished. The large shrub form is of little value in the residential landscape because of its size but it still is useful in the larger landscape of the urban or regional park. The species is highly susceptible to aphid attack.

Lonicera tatarica

Cultivars of *Lonicera tatarica*

'Arnold Red' A compact plant with dark red blossoms and fruit; resistant to aphid attack.

'Beavermor' An outstanding cultivar from Agriculture Canada (Beaverlodge and Morden Research Stations). Bright red flowers edged with pink. Fruit, orange.

'Carleton' Deep pink blossoms.

'Morden Orange' Pink blossoms and orange fruit.

Lonicera tatarica 'Arnold Red'

Lonicera × *xylosteoides*
Vienna Honeysuckle

Size	0.5 –1 m.
Form	Ball-shaped, closed to base.
Flowers	Juvenile plants do not produce flowers, but ccasionally a few yellowish flowers will appear on the cultivar 'Clayvey's Dwarf'.
Texture	Medium.
Foliage	Dull green.

This hybrid between *L. tatarica* and *L. xylosteum* has produced two worthwhile cultivars:

Lonicera × *xylosteoides* 'Claveys Dwarf'

Cultivars of *Lonicera* × *xylosteoides*

'Clavey's Dwarf' This is a juvenile plant with a very dense habit of growth. Flowers are rare and not attractive.

'Miniglobe' The miniglobe honeysuckle is a good, small shrub similar to, but better than, 'Clavey's Dwarf.' The plant does not produce flowers. It too is hardier than the parent *L. xylosteum*. It is an excellent plant for use as a neat, low 0.45 to 0.6 m hedge.

Lonicera × *xylosteoides* 'Miniglobe'

Lotus

Trefoil

Lotus is a clump-forming herbaceous legume with bright yellow flowers. The plant is a European native and one that is drought-resistant and salt-tolerant. The plant is fresh and attractive. It has little landscape value on small sites, but could have many uses on larger ones. There are two characteristics that make for easy identification of this plant.

They are:

- The slender 2.5 cm long seed pods which radiate from a single point and resemble a bird's foot
- The compound leaves which bear three leaflets at the tip, and two more at the base of each leaf.

Lotus corniculatus
Bird's-foot Trefoil

Size	0.3 m.
Form	Sprawling-plant with stems up to 0.6 m long.
Texture	Fine.
Foliage	Compound.
Flowers	Solitary or in heads, showy, bright yellow.

This groundcover is useful for roadside banks and for erosion control on rural properties.

Lotus corniculatus

Maackia

Maackia

Small, flat-topped, deciduous trees belonging to the pea family (*Fabaceae*), that are useful because of their late-summer bloom. The form of tree is graceful and the foliage, which is pinnately compound, contributes to this.

Maackia amurensis
Amur Maackia

Size	4 m.
Form	Upright-spreading small tree.
Habit	Decurrent.
Canopy	Open.
Texture	Medium.
Foliage	Green, pinnately compound.
Flowers	White, pea-like, in upright racemes.
Fruit	A one- to five-seeded pod.

This is a rare, interesting, small tree that is fully hardy. Its flowers are showy and produced in early summer.

Maackia amurensis

Mahonia

Oregon-Grape

These broad-leaved evergreen shrubs are useful as foundation planting material and in the shrub border. The leaves of mahonia cease to function during prairie winters but they persist. The chief value of this plant lies in its foliage, though it also produces very attractive flowers and fruit. *Mahonia* belongs to the Barberry family *(Berberidaceae)* and, like some other members, is considered an alternate host to grain rusts; nevertheless, it is still grown as an ornamental.

Mahonia aquifolium

Mahonia aquifolium
Oregon-Grape

Size	0.6 m.
Form	Low-spreading.
Texture	Medium.
Foliage	Green leaves with short holly-like spines on the margins.
Flowers	Yellow, in racemes.
Fruit	Light blue clusters of pea-sized fruit.

The Oregon-grape is a very ornamental shrub because of its flowers, fruit and foliage. Autumn color of foliage is very rich purplish red. Like other broad-leaved evergreens used in the region, these plants must not be grown in locations where the foliage will be exposed to the strong sunlight of early spring. These conditions will result in severe burning and loss of leaves.

Mahonia repens

Mahonia repens
Creeping Oregon-Grape

Size	0.1–0.15 m.
Form	Low, spreading groundcover.
Texture	Medium.
Foliage	Dull green with spiny margins.
Flowers	Yellow, in racemes.
Fruit	Pea-sized, covered with light blue blush.

This is not as attractive as *M. aquifolium* because it lacks the lustrous foliage. It is, nevertheless, a worthy groundcover for shady areas.

Malus

Crab Apple

These small, dense, low-headed trees are relatively short-lived in the prairie provinces when not well cared for but with annual pruning and good management they can be expected to last for 40-50 years. The crab apples are among the showiest of all woody plants used in the landscape. Their flowers, which appear soon after leaf emergence, are produced in great abundance and come in white, deep pink, or purplish red. While the leaves of most species are green, there are cultivars with red leaves and red-veined leaves. The fruit is a five-celled pome with great variation in size. Species and cultivars vary greatly in hardiness and in resistance to the devastating bacterial disease fire-blight. Crab apples prefer deep, well-drained soils and sheltered sites.

Malus arnoldiana 'Tanner'
Tanner Crab Apple

Size	3 m.
Form	Low-headed, upright- spreading.
Habit	Decurrent.
Canopy	Dense.
Texture	Coarse.
Foliage	Green.
Flowers	White, produced in great abundance.

A cultivar with white fragrant blossoms, produced in abundance at only three years of age.

Malus arnoldiana 'Tanner'

Malus × *adstringens* Cultivars
Rosybloom Crab Apples

'Almey' A tree with an open configuration. Flowers are large, deep rose in color. Each petal has a white base. Fruit is small, bright red and retained over winter.

'Arctic Dawn' The hardiest of all the rosyblooms; also resistant to fire-blight.

'Hopa' A tree with a wide-spreading open configuration. Flowers are a purplish pink, fading to a dull mauve. Fruits are small, red and dropped early, creating some maintenance problems.

'Kelsey' This is the first semi-double flowered rosybloom produced on the prairies. It is a spur-type plant with the flowers held closely to the branches. Blossoms are 5 cm in diameter and are purplish red. Each petal has a white marking at its base. Foliage is red changing to coppery-green with red veins.

Malus × *adstringens* 'Almey'

Malus × *adstringens* 'Arctic Dawn'

'Pygmy' This is a very twiggy, dense dwarf tree with a tight globular form to 4 m. Foliage is dark purplish green, and flowers are deep purplish red. The plant rarely blooms until it is 8–10 years old. It is a useful plant where a tree with a formal, sheared look is required.

'Red Splendor' A small, rounded tree with deep red buds. Flowers are light pink. The small scarlet fruit, which is retained over the winter, contributes to the ornamental value of the tree.

'Royalty' This tree is tall and upright to 6 m, with outstanding deep purple foliage. It is susceptible to fire-blight and cannot be recommended for this reason.

'Rudolph' This is a small, round-topped tree with dark red flower buds and deep rose flowers.

'Selkirk' A small, pink flowering tree with good form and bright red, persistent fruit.

'Strathmore' This tree grows to 5 m with an upright-form, bronzy foliage, and pink flowers.

'Thunderchild' This cultivar is a popular small, deep purple leaved tree with a round, spreading head. The flowers are pink and contrast nicely with the immature foliage. Fruit is small, flattened, and purplish red. It is more resistant to fire-blight than 'Royalty' and therefore is preferred when a type with deep purple foliage is called for..

Malus baccata

Malus baccata
Siberian Crab Apple

Size	5 m.
Form	Low-headed, ball-shaped.
Habit	Decurrent.
Canopy	Dense.
Texture	Coarse.
Foliage	Green.
Flowers	White, very showy.
Fruit	Small, bright red, with deciduous calyx.

This is a very hardy spring-flowering tree which has played a part in the early development of crabapple and apple-crab hybrids for the prairie provinces. Susceptibility of the various cultivars to the bacterial disease fire-blight varies from very susceptible to highly resistant, so caution is advised.

Cultivars of *Malus baccata*

'Columnaris' This is a very narrow, columnar, flowering tree to 5 m. It may be useful for accent purposes and for those locations where space for tree planting is limited but it is hardly a thing of beauty. However the small red fruit is persistent and quite decorative.

'Snowcap' A highly recommended cultivar from the Beaverlodge Research Station. Winter effect of its small, red, persistent fruit is its outstanding characteristic.

Malus baccata 'Columnaris'

Malus × purpurea 'Red Jade'
Red Jade Crab Apple

Size	1–3 m.
Form	Small weeping tree.
Habit	Decurrent.
Canopy	Open.
Texture	Coarse.
Foliage	Bright green.
Flowers	Pink.
Fruit	Small, red, cherry-like.

This small tree with its pendulous branches can have high value as a specimen in some locations. Very well suited to rockeries and oriental style landscapes. It is attractive in the fall when the fruit is changing from yellow to red. Some training is required for the plant to grow upright.

Malus × purpurea 'Red Jade'

Ononis
Ononis

These are attractive but rare, small, neat woody plants from southern and central Europe. The genus is a member of the pea family *(Fabaceae)*. The one species described grows in full sun on well-drained soils and will tolerate some salinity. It is one of the better small, flowering shrubs suited to the scale of the home landscape.

Ononis spinosa
Rest Harrow

Size	0.6 m.
Form	Globular.
Texture	Fine.
Foliage	Light green, compound in 3's (trifoliate).
Flowers	Pink.

This neat, small, spiny, summer-blooming plant thrives in the drier parts of the region.

Ononis spinosa

Pachysandra

Spurge

Vigorous, low-growing, yet upright, leafy groundcover plants with matted rootstocks. Although the inflorescence of *Pachysandra* is not showy, it is interesting in that it is a spike with both staminate and pistillate flowers. Staminate flowers occupy the upper portions of the structure and, following fertilization, they disintegrate leaving the pistillate flowers to develop fruit. The fruit is a white berry equal in size to that of the saskatoon.

Pachysandra terminalis
Japanese Spurge

Size	20-30 cm.
Form	Upright, leafy plant that spreads by rhizomes or underground stems.
Texture	Coarse.
Foliage	Lustrous dark green, persistent.
Flowers	White, imperfect/monoecious, borne terminally, not outstanding.

This is a hardy, superior groundcover for deep and light shade. It will not do well in sunny locations. Like most groundcovers, a deep soil high in organic matter is preferred.

Pachysandra terminalis

Parthenocissus

Virginia Creeper

These plants belong to the grape family *(Vitaceae)* and are among the region's most reliable climbers. The leaves are large, coarse, and palmately compound with up to five leaflets per leaf. Fruit is a small blue berry, borne in grape-like clusters. The value of these plants lies in their ability to climb and spread.

Parthenocissus quinquefolia
Virginia Creeper

Height	30 m.
Form	Climber.
Texture	Coarse.
Foliage	Green; leaves with five leaflets.
Flowers	Inconspicuous, in terminal cymes.
Fruit	Bluish-black berry.

This plant requires support since it climbs by tendrils that wrap themselves around anything that is handy. The leaves of plants growing in full sun generally color up nicely in the fall of the year. The plant has been used as a groundcover on banks, but requires some maintenance. It is very susceptible to powdery mildew in places where there is little air movement and some shade.

Parthenocissus quinquefolia

Parthenocissus quinquefolia var. *engelmannii*
Self-Clinging Virginia Creeper

Height	30 m.
Form	Climber.
Texture	Coarse.
Foliage	Similar to the species but slightly smaller.
Flowers	In terminal cymes.
Fruit	Bluish-black berry.

Like the species, this plant will show spectacular red autumn coloration when grown in full sun. Unlike the species the variety has adhesive pads which fasten themselves to vertical surfaces; hence it cannot be recommended for growing against wooden structures that will require regular maintenance. The variety is also not as subject to powdery mildew as the species.

Parthenocissus quinquefolia var. *engelmannii*

Paxistima

Cliff-Green, Oregon-Boxwood

Low, broad-leaved evergreens suitable for underplanting. Both species tend to be injured when exposed to strong, early spring sunlight. Although cliff-green prefers a well-drained soil and a location that is not in deep shade, the Oregon-boxwood prefers a shady environment.

Paxistima canbyi

Paxistima canbyi
Cliff-Green

Size	0.2 m.
Form	Dense, upright plant.
Texture	Fine.
Foliage	Small, somewhat leathery, green leaves with toothed margins.

Cliff-green is a good groundcover for semi-shade. When planted closely, on 10–15 cm centers, it will compete favorably with weeds. This plant spreads by underground rootstocks. It seems to do well in most environments, but will survive the winter better if it has the protection of persistent snow cover.

Paxistima myrsinites

Paxistima myrsinites
Oregon-Boxwood

Size	0.45 m.
Form	Mound-like.
Texture	Fine.
Foliage	Dark, glossy green.

This is a very attractive small shrub or groundcover for shady areas. It has been seen growing in sunny spots; however, it must be assured protection from early spring sunshine to avoid "burning" of the foliage. If snow cover is retained in these situations, there is generally no problem.

Phellodendron

Corktree

Phellodendrons are small to medium-sized, high-headed deciduous trees, native to eastern Asia. They have attractive, large, compound leaves and a thick bark that is soft and corky. The flowers are dioecious, small, yellowish-green, and borne terminally. The species described is highly recommended.

Phellodendron amurense
Amur Corktree

Size	9–10 m.
Form	High-headed, upright-spreading.
Habit	Decurrent.
Canopy	Open.
Texture	Coarse.
Foliage	Light green, compound.
Bark	Corky, spongy to touch.
Fruit	About the size of a pea, berry-like; gives off a turpentine-like odor when bruised.

A neat, small tree with good, soft, yellow color in the fall. Subject to root injury from low temperatures in the north but seems to be quite satisfactory elsewhere.

Phellodendron amurense

Philadelphus

Mockorange

These flowering shrubs usually produce fragrant white blossoms in early summer. Several double-flowered cultivars are available. The mockorange generally, is a stiff, upright, somewhat leggy shrub that has little to commend it once the flowering period is over. Because of this shortcoming it is a plant that is suited more to the larger landscape than to that of the average residential property. Flowers are borne on lateral branches usually in racemes, but sometimes solitary or in two- to three-flowered cymes.

Philadelphus coronarius

Philadelphus coronarius
Sweet Mockorange

Size	3 m.
Form	Upright.
Texture	Coarse.
Foliage	Light green.
Flowers	White, fragrant.

The sweet mockorange is a stiff, leggy, upright shrub. It is rare, but has been frequently used as parental material in the development of new cultivars. It tolerates dry situations.

Philadelphus lewisii 'Minnesota Snowflake'

Philadelphus lewisii 'Waterton'
Waterton Mockorange

Size	1–2 m.
Form	Globular.
Texture	Medium.
Foliage	Green.
Flowers	White, scentless in 5–9-flowered racemes.

This is a selection from plants found in Waterton National Park. The cultivar blooms freely in early summer and carries its blossoms right to the base of the plant. It does best where supplementary moisture is available. In northern areas, plants may lose flower buds on shoots that extend above the snow line.

Philadelphus lewisii 'Waterton'

Hybrids of *Philadelphus*

The cultivars worthy of trial in the prairie region are all hybrids.

× **'Audrey'** An Agriculture Canada, Morden introduction, 2.5 m shrub with large white fragrant blossoms.

× **'Marjorie'** (also a Morden introduction) Taller than 'Audrey' and of different parentage though flowering habit is similar. It has arching branches.

× **'Minnesota Snowflake'** A tall, leggy shrub producing an abundance of pure white, starburst or water-lily type blossoms.

× **'Patricia'** To 1 m with creamy white blossoms.

× **'Purity'** To 1.5 m with fragrant porcelain-white blossoms. May not flower annually.

× **'Sylvia'** A 2 m tall, upright shrub with semi-double fragrant flowers.

Phlox

Groundcover Phlox

These low, mat-forming herbaceous perennials are vigorous, colorful, spring-blooming subjects. There are several species and cultivars available and all are attractive to look at but not all of them have the substance required of good groundcover material. The one species listed is by far the best groundcover phlox because it not only provides a long-lasting, brilliant display of color in the spring, but also a dense, tough carpet of green for the rest of the season.

Phlox borealis
Arctic Phlox

Size	Less than 0.15 m.
Form	Low, ground-hugging, somewhat hummocky, foliage is persistent.
Texture	Fine.
Foliage	Dark green during the growing season, but changing to a bleached yellow when the cold autumn weather arrives.
Flowers	Bright pink and produced in such great abundance they literally cover the foliage.

This species is by far the best *Phlox* for groundcover purposes.

Phlox borealis

Physocarpus

Ninebark

These are vigorous, coarse-textured shrubs of upright-spreading habit. The bark is thin and golden brown, peeling in sheets. The leaves are three-lobed, rounded at the tips, and deeply veined. Flowers are usually white, somewhat conspicuous and produced in terminal umbel-like racemes in early summer. The inflated pods are more conspicuous than the flowers and quite attractive. Leaves are alternate.

Physocarpus opulifolius

Physocarpus opulifolius
Common Ninebark

Size	2–3 m.
Form	Upright-spreading.
Texture	Coarse.
Foliage	Green.
Flowers	Creamy white.
Bark	Cinnamon brown exfoliating bark.
Fruit	A dry capsule. The clusters turn red as they mature and are actually more attractive and showier than the flowers.

Ninebark is adapted to most situations. It has been observed that weeds fail to do well in soil beneath the canopy of this plant. It has not been established whether this phenomenon is induced by shade or by some substance released by the plant.

Cultivars of *Physocarpus opulifolius*

'Dart's Gold' A smaller selection of 'Luteus' with intense, golden foliage.

'Luteus' (golden ninebark) A form of the species with greenish yellow foliage. The golden-yellow color is intensified where plants grow in full sunlight.

Physocarpus opulifolius 'Darts Gold'

Picea

Spruce

These coniferous trees are usually of pyramidal habit. Their form and density is carried throughout life. Branches are whorled; branchlets have prominent leaf cushions which are projected into peg-like stalks which bear the individual needles. Buds are ovoid with or without resin. Leaves are spirally arranged needles. Flowers (cones) are imperfect/monoecious. The female or seed cones are borne terminally in the upper part of the tree. Male cones are small, yellowish or reddish, and are borne laterally in the lower part of the tree. Most frequently seen pests are the Cooley spruce gall aphid, the red spider mite and the spruce budworm.

Picea abies
Norway Spruce

Size	15 m.
Form	Narrow pyramid.
Foliage	The dark green needles in this species are distinctive. They are four-sided in cross-section but, unlike other species discussed, the form is trapezoidal rather than square. Needles with a square cross-section can be rolled around easily between thumb and forefinger but the needles from the Norway spruce cannot. The secondary branches of this species are pendulous, giving the tree an appearance quite different from that of other species.
Bark	Reddish brown.

Seed Cones Pendulous, cylindric 10–15 cm.

This tree seems well adapted to most regional conditions. The dense dark green of the foliage and the pendant secondary branches tend to give Norway spruce its unique landscape qualities.

Picea abies

Cultivars of *Picea abies*

'Compacta' This small, compact, conical plant will retain its density and form without any shearing.

'Gregoryana' A very slow-growing, globular plant that is very suitable for use in rockeries. It prefers a well-drained, sheltered location in full sun, but may require some shade in early spring.

'Nidiformis' (nest spruce) This is a somewhat prostrate plant. The semi-erect branches curve outwards rather than to the center of the plant, leaving a small, centrally-located depression, hence the common name.

Picea abies 'Nidiformis'

Picea engelmannii
Engelman Spruce

Size	30 m.
Form	Narrow pyramid.
Foliage	The bluish green needles are much like those of white spruce but give off a distinctive odor when crushed.
Bark	Scaly, coarse textured, orangish brown.
Seed Cones	Cylindric, 7 cm with loose fitting scales.

Young trees differ little from white spruce in a landscape sense; hence the species is not carried by nurseries.

Picea engelmannii

Woody Ornamentals

Picea glauca

Picea glauca
White Spruce

Size	20 m.
Form	Narrow pyramid.
Foliage	Needles are dull, dark green, sometimes blue-green.

Seed Cones 5 cm, cylindrical; cone scales have entire margins.

The white spruce is commonly found on upland sites in the north central and northern parts of the region. In the grasslands to the south, it will be found growing only in valley bottoms and on north-facing slopes. It is a good landscape tree because of form, density and year round appearance. It is effective as an individual specimen or in groups.

Cultivars of *Picea glauca*

'Densata' A more compact form of the species known as the Black Hills spruce.

'Albertiana Conica' This is a popular, small, true-dwarf tree. Culture has been rather difficult due to its susceptibility to both direct and reflected light in the early spring. The plant appears to be hardy in the north central part of the region, but favorable micro-climates are important.

Picea glauca 'Albertiana Conica'

Picea omorika 'Nana'

Picea omorika
Serbian Spruce

Size	15 m.
Form	Narrow pyramid with short ascending branches.
Foliage	This tree has a frosty look due to two broad white bands on the underside of the needles. Young needles turn upwards so the "frostiness" is more commonly seen on new growth.

Seed Cones Lustrous brown, 3–6 cm ovoid-oblong.

This tree prefers sheltered locations where supplementary moisture is available. It is slow growing.

Cultivars of *Picea omorika*

'Nana' An attractive dwarf pyramid for shady sites and rich, well-drained soils.

Picea pungens
Colorado Spruce

Size	20 m.
Form	Tall, narrow pyramid with whorls of horizontally arranged branches.
Foliage	Color varies from green to blue to silvery green. Many cultivars have been selected on the basis of foliage color. Needles are squarish in cross-section and very sharp-pointed.

Seed Cones 7 cm long.

This tree has no special cultural requirements. It is fully hardy. Blue and silvery blue forms are very popular and are commonly used as single specimens. They are also very effective in mass planting. This is probably our most drought-tolerant spruce.

Picea pungens 'Glauca globosa'

Cultivars of *Picea pungens*

'Globosa' This small, dense, globular shrub has the capability of maintaining its natural form without assistance.

'Procumbens' The good, bright blue foliage color is the main ornamental feature of this plant. Because of its form and stature, its use is limited.

'Hoopsii' A dense narrow pyramid. It has a good silvery blue color that is carried right to the trunk.

'Koster' An old cultivar with silvery white foliage.

'Moerheimii' A narrow compact pyramid with strikingly blue foliage.

'Montgomery' A compact form to 3 m. It carries its blue color well in towards the trunk of the tree.

'Morden Blue' Dense pyramidal tree with an overlapping branching habit.

Picea pungens 'Glauca procumbens'

Picea pungens 'Montgomery'

Woody Ornamentals

Pinus

Pine

Pine trees are conifers with whorled branches. Needles, which are borne in bundles of two, three, or five, are spirally arranged about the branchlets and twigs. Trees invariably have a dense pyramidal or columnar form when young but almost all have the tendency to self prune their lower branches, thus exposing trunks as they get older. Older trees also have a tendency to retain needles on the tips of the branches and shed those closer to the trunk.

Flowers (cones) are imperfect/monoecious. The male cones are clustered at the base of the young shoots and disintegrate soon after shedding their pollen. The female or seed cones are lateral or sub-terminal and are much larger than those of the male cones. In most cases seed cones mature the second autumn, releasing the seed in the spring of the third year..

Pines grow naturally in light, sandy soils but will adapt to the heavier soils of the prairies. In transplanting pines to new sites, special care must be taken to assure that the soil ball around the roots is not broken, otherwise they will not usually survive the move.

Pinus albicaulis

Pinus albicaulis
Whitebark Pine

Size	10 m.
Form	Upright-spreading crown.
Foliage	Needles five to a bundle, stiff, bluish green, 3–8 cm long.
Bark	On young stems the bark is chalk white and smooth, becoming rough with brown scaly plates.
Seed Cones	Egg-shaped, without prickles, 8 cm; fruit is nut-like, the size of a small pea.

Sunny, well-drained sites are preferred. This is an attractive tree but it is rarely seen in cultivation.

Pinus aristata

Pinus aristata
Bristlecone Pine

Size	4 m.
Form	Twisted, irregular.
Texture	Fine.
Foliage	Dark green, flecked with a distinct white exudate, five needles to a bundle.

This is a shrubby tree with distinctive arm-like limbs covered with whorls of needles. It is an excellent plant for the rock garden or for those landscapes where a plant with a "weathered" look is called for.

Pinus banksiana
Jack Pine

Size	20 m.
Form	Broad-oval with slender, twisted branches.
Foliage	Needles to 3 cm, in bundles of two, dull, yellowish green.
Bark	Dark brown, scaly.
Seed Cones	Claw-shaped, 2.5–5 cm, yellowish, smooth, curved. Cone tips point forward to the ends of the branches.

There are no special cultural requirements. It is a native tree but not widely used as an ornamental and seldom carried by nurserymen. The twisted form of the Jack pine can be interesting; however, the Scots pine is much preferred when this tree form is called for.

Pinus banksiana

Pinus cembra
Swiss Stone Pine

Size	15 m.
Form	Columnar to pyramidal.
Foliage	Needles are soft, fine-textured dark green, in bundles of five. The shoots are tomentose with thick brown hairs. This is a very dense tree. It retains foliage on lower branches much longer than other members of the genus.
Bark	Smooth, greyish.
Seed Cones	Globular 5–8 cm, without prickles; seed a pea-sized nut without wings. The seeds are edible as "pine nuts".

There are no special cultural requirements. This is a very attractive conifer. Its foliage is very resistant to sunscald injury. This tree has been much under-utilized. It is highly recommended.

Pinus cembra

Pinus contorta var. *latifolia*
Lodgepole Pine

Size	25 m.
Form	Tall, straight-trunked, narrow.
Foliage	Needles 5–7 cm, dull, yellowish green in bundles of two.
Bark	Rough, brown and scaly.
Seed Cones	Claw-shaped, 2.5-5 cm. Each cone scale is armed with a small, sharp but fragile prickle. Cone tips point towards the trunk of the tree.

This is not one of the better conifers for landscape purposes although it is frequently used in farm shelterbelts.

Pinus contorta var. *latifolia*

Woody Ornamentals

Pinus flexilis

Pinus flexilis
Limber Pine

Size	8 m.
Form	Short-trunked with a large uneven crown. Often seen in multi-stemmed form.
Foliage	Needles, five to a bundle, 3–8 cm bluish green.
Bark	Grey, smooth on young trees becoming dark brown and scaly with age.
Seed Cones	Cylindrical, 6–16 cm, short-stalked, not pendulous, but arising in all directions from the tips of branches.

Limber pine is rarely seen in cultivation though it has some value as a shrub-like tree. It is often mistaken for *Pinus albicaulis*; however, it can be easily identified when seed cones are present.

Pinus mugo

Pinus mugo
Mugo Pine, Swiss Mountain Pine

Size	Variable, 0.6–3 m.
Form	Commonly either low mound-like or upright globular. A columnar form also exists, but is not common.
Texture	Medium.
Foliage	Dense, dark green. Two needles to a bundle.

The species is quite variable. Botanists recognize three botanical varieties: *pumilio, rostrata,* and *rotundata.* The true dwarf form (var. *pumilio*) is much sought after and yet it is difficult to obtain. Usually the customer ends up with a plant that soon loses its dwarfness. Annual shearing at the time of new growth will help for a few years but eventually such specimens will lose their natural look.

Cultivars of *Pinus mugo*

'Compacta' A very dense, globular shrub to 1 m.

'Teeny' A small, neat, short-needled plant with a dense globular habit.

Pinus mugo var. *pumilio*

Pinus nigra
Austrian Pine

Size 15 m.

Form Upright-spreading with a large crown.

Foliage Needles are sharp-pointed, rigid, two to a bundle, 15 cm long, and dark glossy green.

Seed Cones 7.5 cm.

At maturity this is a picturesque, large, flat-topped tree with an open crown.

Pinus nigra

Pinus ponderosa
Western Yellow Pine, Ponderosa Pine

Size 15 m.

Form Stoutly cylindrical; in later life it self prunes, exposing a strong straight trunk. Lower branches have a tendency to hang down.

Foliage Dull, yellowish green, needles, 15–20 cm long, stiff, sharp-pointed, three to a bundle.

Bark Orange-brown, furrowed.

Seed Cones Pyramidal shape, 7.5 cm in diameter at the base, each scale armed with one sharp prickle.

Pinus ponderosa

This is a good tree for park-like surroundings where it can be used in mass. Individual specimens of mature trees not as useful, though young trees can be quite attractive. It is very drought-resistant.

Variants of *Pinus ponderosa*

var. *scopulorum* A smaller, more compact form which is native to the Black Hills of South Dakota.

Pinus resinosa
Red Pine, Norway Pine

Size 15 m.

Form Upright-oval, self prunes to a short, somewhat open crown and long straight trunk; whorls of branches very distinct.

Foliage Green, needles in bundles of two, straight, 5–8 cm, coarse.

Seed Cones Conical, 3–6 cm, scales thick, unarmed.

At maturity this is an attractive tall conifer of the same landscape type as the eastern and western white pines and the Douglas-fir.

Pinus resinosa

Woody Ornamentals

Pinus strobus

Pinus strobus
Eastern White Pine

Size	15 m.
Form	Upright oval, develops an asymmetric branching habit at maturity. Lower branches tend to grow horizontally while those in the upper part of the tree ascend more or less vertically. Trunks of mature trees are very tall and straight.
Foliage	Needles in bundles of five, 5–10 cm long, soft, straight, flexible, light blue-green.
Bark	On young trees, smooth, grey-green; on mature trees, dark greyish brown with deep longitudinal furrows.

Seed Cones 10–15 cm, smooth, cylindrical, pendulous.

The tall, majestic form of mature trees is attractive, as is the somewhat dense, fine, soft texture of young trees. Foliage is very susceptible to browning and needle-cast in late winter of establishment years.

Pinus sylvestris

Pinus sylvestris 'Fastigiata'

Pinus sylvestris
Scot's Pine

Size	15 m.
Form	Pyramidal when young with branches to the ground. Mature form lacks the symmetry of youth and is replaced with a somewhat picturesque form with a broad, open crown.
Foliage	Needles 5–7 cm, twisted, blue-green in bundles of two.
Bark	That of the trunk is brown, but branches and trunk in the upper parts of the tree tend to shed the outer bark, exposing an attractive light orange inner bark. When dead or weak branches are removed, this feature becomes especially striking.

Seed Cones To 5 cm, green, curved and tightly fastened to stems. Cone scales without prickles.

In its mature form this tree is without question the best of the pines for use in the urban landscape.

Cultivars of *Pinus sylvestris*

'Fastigiata' (sentinel pine) This is a very narrow, small, accent-type tree for special effects. It is particularly well-suited to the landscape of the rock garden.

'Viridis Compacta' (green compact Scot's pine) This is a very compact, dwarf tree resembling very much the small pine trees that are harvested and used at Christmas time.

Polygonum

Fleeceflower

A genus of small-statured plants, consisting of herbs, climbers and sub-shrubs. The stems of the plants are erect and have swollen nodes, covered by a membranous sheath in some cases. The inflorescence is axillary and consists of clusters of several small flowers at each node.

The leaves are alternate, leathery or somewhat fleshy. The one species discussed here is a useful groundcover and not in any way like the many fleeceflowers that are sold and used as herbaceous perennials.

Polygonum cuspidatum 'Compactum'
Compact Japanese Fleeceflower

Size	0.4 m.
Form	Upright.
Texture	Coarse.
Foliage	Green with red veins and petioles.
Fruit	Dry, compressed capsules in terminal spikes, reddish and attractive in early autumn.

Stems of this plant develop from deep wine-red annual shoots. An excellent tall groundcover for full sun or partial shade. Because of its height, it would not be suitable for all situations. This plant was formerly known as *Polygonum reynowtria*.

Polygonum cuspidatum 'Compactum'

Populus

Poplar

This group of trees is quick-growing but short-lived. They are not good landscape trees because of their many undesirable characteristics. They are shallow-rooted and their roots when injured will respond by producing sucker shoots. This response, however, is almost minor compared to that which can occur when a mature tree is cut down. Such trees have a very extensive root system and it is not uncommon in urban areas to find, in response to the removal of one tree, that sucker shoots will emerge on adjoining properties as well as on the one on which the tree was growing. Another objectionable feature of the poplar is the seed produced. Seeds are appended by a tuft of fluffy white hairs and this "poplar-fluff", which in many instances can pile up like snow, creates a variety of problems. Poplars are dioecious, hence it is the female trees that are bothersome in this case. People have learned to avoid planting female clones of poplar, yet such trees in native stands continue to make their presence felt each spring.

Poplars also are subject to canker injury. Cankers which are first recognized as swellings on the trunk are primarily a response to some form of injury (mechanical or frost) following which the organism responsible for the canker invades the tissue. As the disease advances, tissues will split and exude an amber colored liquid. As the canker develops, further trunk disintegration occurs leading to death or to breakage of the tree at the infection site.

Populus alba 'Nivea'

Populus alba 'Nivea'
Silver Poplar

Size	9 m.
Form	Upright-spreading.
Habit	Decurrent.
Canopy	Open.
Bark	The upper trunk is covered with conspicuous small lenticels.
Texture	Coarse.
Foliage	Three-lobed leaves, glossy green above, intense silvery white beneath.

This tree is very shallow-rooted and it has a strong suckering tendency. Mature trees are quite attractive but the associated problems can make their value questionable. The species is not fully hardy. Because of its three-lobed leaves, it is often mistaken for maple. However, with maple, the leaves are opposite one another while on the poplar they are arranged alternately.

Populus balsamifera

Populus balsamifera
Balsam Poplar

Size	30 m.
Form	Upright oval.
Habit	Decurrent.
Canopy	Open.
Texture	Coarse.
Foliage	Deep, glossy green. Buds large, sticky.
Bark	Grey-green, deeply furrowed in mature trees.

This is the large native poplar commonly seen in the river valleys in the north-central and northern parts of the region. While the forest tent caterpillar *(Malacosoma spp.)* seems to prefer the foliage of native poplars, the balsam poplar is not subject to attack by this insect, unless food supplies have been exhausted.

Populus berolinensis

Populus × *berolinensis*
Berlin Poplar

Size	12 m.
Form	Upright oval.
Habit	Excurrent.
Canopy	Dense.
Texture	Coarse.
Foliage	Ovate to oval, glossy green.

This is a male clone with a rather neat, upright habit which is generally maintained for at least 20 years.

Populus × *canescens* 'Tower'
Tower Poplar

Size	8–10 m.
Form	Columnar.
Habit	Excurrent.
Canopy	Dense.
Texture	Medium.
Foliage	Green leaves with wavy margins and a pointed tip; undersides are silvery green due to numerous fine hairs.

'Tower' is the best columnar poplar for the region. It is a neat, vigorous tree. Root systems from container-grown stock are far superior to those from field grown plants; hence, when transplanting, use of the latter should be avoided. In spite of its good columnar habit, the root sytem of 'Tower' can be expected to be as widespread and as bothersome as that of any other poplar.

Populus × *canescens* 'Tower'

Populus deltoides
Plains' Cottonwood

Size	30 m.
Form	High-headed, with broad canopy.
Habit	Decurrent.
Canopy	Dense.
Texture	Coarse.
Foliage	Green, deltoid leaves.

This is a handsome large tree, native to the watercourses of the south. It rivals the American elm in both size and stature. It has been used successfully as a street tree in the southern parts of the region.

Populus deltoides

Populus × *jackii* 'Northwest'
Northwest Poplar

Size	15 m.
Form	Broad, upright oval.
Habit	Decurrent.
Canopy	Closed.
Texture	Coarse.
Foliage	Green with flattened base.

This male clone is very common in the region even though specimens growing in the north central part appear highly susceptible to canker. It is, however, resistant to leaf rust. A good shelterbelt tree for the farmstead but too large and with too many problems for use on urban properties.

Populus × *jackii* 'Northwest'

Woody Ornamentals

Populus × petrowskyana

Populus × petrowskyana
Russian Poplar

Size	15 m.
Form	Upright pyramidal.
Habit	Decurrent.
Canopy	Dense.
Texture	Coarse.
Foliage	Green, oval with a wedge-shaped base.

This is one of the early hybrid poplars used in western Canada as a street tree. It is a female clone and produces much "fluff". It also had a bad reputation involving roots and sewer lines but it is likely no worse than any of the others in this respect.

Cultivars of *Populus × petrowskyana*

Populus × petrowskyana 'Griffin'

'Dunlop' This is a straight, strong, vertical-growing poplar. It is resistant to leaf rust but susceptible to canker. It has been suggested that the cultivar is dioecious but a more likely explanation has been put forward suggesting that because some plants did not produce seed until quite mature the notion had arisen that some plants were male.

'Griffin' This is a hybrid between *P. deltoides* and *P. × petrowskyana*. It is a male clone of narrow pyramidal form. During its youth it carries its branches low to the ground but becomes high-headed in later life. In some areas this tree is more susceptible to *Septoria* canker than it is in others. This may well be an indication of the fact that the tree is more subject to a primary environmental stress in some parts of the region than it is in others. For instance, *Septoria* canker is considered a secondary invader of frost canker injury which in itself is an indication that the injured plant is not hardy.

'Walker' This vigorous, narrow- crowned hybrid is resistant to rust. It is a good shelterbelt tree, but it is a female clone.

Populus tremula 'Erecta'
European Columnar Aspen

Size	9 m.
Form	Columnar.
Habit	Excurrent.
Canopy	Dense.
Texture	Medium.
Foliage	Green both sides, circular, with wavy margins.

This tree has two characteristics which tend to detract from its good columnar form. The first is its tendency to have the lower 30–60 cm of the trunk devoid of foliage. The second is the appearance of the uppermost branches which tend to lack the stiff upright habit of lower branches.

Populus tremula 'Erecta'

Populus tremuloides
Trembling Aspen

Size	25 m.
Form	Upright oval, high-headed.
Habit	Decurrent.
Canopy	Open.
Texture	Medium.
Foliage	Light green, circular with flattened petioles.
Bark	Smooth greenish-white to almost white in northern parts of the region.

This is a very common native tree occupying upland areas in the landscape of the parkland and of the boreal forest. It is susceptible to canker and forest tent caterpillar *(Malacosoma spp.)*. The stature of trees in the aspen parkland is much shorter than those of the boreal forest. Bright yellow and sometimes orange fall color is the outstanding feature of this species. Throughout the region, the extent and limits of male and female clones become evident in early summer when the female clones are producing seed. Aspen poplar is very sensitive to environmental change. It has been noted that when recreational use is permitted beneath trees to the extent that the understory is stressed or destroyed, then decline of the aspen soon follows.

Populus tremuloides

Populus tremuloides

Woody Ornamentals

Potentilla

Cinquefoil, Potentilla

The cinquefoils are chiefly small deciduous flowering shrubs, occasionally groundcovers, with fine-textured compound pinnate leaves. Flower color is variable: yellow, orange, white and pink-flowered cultivars are common. Flowers are single, although semi-double forms are now available. Flowers are borne over a long season.

The potentilla is circumpolar in its distribution. In western Canada the shrubby forms are more commonly found in areas of high rainfall, that is, in the foothills and the mountains.

Potentilla fruticosa

Potentilla fruticosa 'Katherine Dykes'

Potentilla fruticosa
Shrubby Cinquefoil

Size	0.2 -1.5m
Form	Globose, closed to the base.
Texture	Fine.
Foliage	Green.
Flowers	Bright yellow.
Bark	Rough, exfoliating, not attractive.

This is a small compact shrub with a very long flowering season. Winter effect is poor because of its untidy appearance after leaf-drop, with many leaves and dead flowers being retained. It is hardy in all parts of the region and adapted to all but the very driest sites. It has a preference for sunny locations. The shrub has no serious pests.

Cultivars of *Potentilla fruticosa*

'Abbotswood' A neat, clean-looking cultivar with spreading habit. Blossoms are very white and are well distributed over the plant. It blooms over a long season.

'Coronation Triumph' A loose plant with bright yellow flowers produced over a very long season. Sepals are prominent between the petals.

'Forestii' A low, spreading, mound-like plant with large yellow flowers produced in abundance. This plant could be used as a good flowering groundcover.

'Goldfinger' A compact plant with good sized, bright yellow flowers.

'Jackmanii' A very upright plant with larger leaves than most other cultivars. Flowers yellow and borne in dense clusters.

'Katherine Dykes' A plant with an arching habit of growth. Flowers are light yellow and distributed along the arching stems. They are borne over a long season.

'Parvifolia' A dense small-leaved, mound form with small yellow flowers and very fine-texured foliage.

'Red Ace' A small, bushy plant with reddish flowers fading to orange. It is not hardy in all parts of the region.

'Snow Bird' Neat, semi-double, white. A University of Manitoba introduction.

'Tangerine' A spreading, mound-forming plant with coppery-yellow flowers. Flower production is sparse compared to most other cultivars though the flowering season is long.

'Yellow Bird' Semi-double yellow from the University of Manitoba.

Potentilla fruticosa 'Red Ace'

Potentilla tridentata
Three-Toothed Cinquefoil

Size	0.2 m.
Form	Spreading.
Texture	Fine.
Foliage	Glossy, dark green.
Flowers	White.

A low groundcover for full sun. Somewhat untidy as flowers fade; requires a lot of dead-heading to keep it neat.

Potentilla tridentata

Prinsepia

Prinsepia

The single species of this genus is a stout, spiny-shrub with arching branches that is well suited for use as a barrier plant. Because of its density and stout spines it is capable of discouraging trespass by both animals and humans. The fruit is a juicy red cherry-like drupe which also gives it some value as a wildlife plant.

Prinsepia sinensis

Prinsepia sinensis
Cherry Prinsepia

Size	1.5 m.
Form	A large coarse-looking shrub with stiff arching branches.
Texture	Medium.
Foliage	Bright green, long and narrow.
Flowers	Small, yellow, not conspicuous.
Fruit	Scarlet, circular but flattened, not globose, about 2 cm in diameter.

This is a very well-armed shrub that can be used to create an impenetrable barrier. Leaves emerge very early in the spring. It has no particular cultural requirements and should be hardy in all parts of the region.

Prunus

Plum, Cherry, Chokecherry, Almond

This group of trees and shrubs contains many worthwhile ornamentals and fruit plants. Leaves are alternate, serrate, rarely entire. Stipules are separate from the petiole and are glandular. The stipules are frequently early deciduous. Flowers are perfect and are either solitary, in dense clusters or in racemes. Petals are white, pink or red.

Flowers appear with, or before the leaves. All fruits are edible when cooked, but many species are too astringent to eat raw. The fruits of these plants are drupes and the seeds inside some pits may be poisonous due to high concentrations of *amygdalin*.

Prunus besseyi

Prunus besseyi
Western Sandcherry

Size	0.5 –1 m.
Form	Upright-spreading.
Texture	Medium.
Foliage	Green.
Flowers	White, not conspicuous.
Fruit	A black cherry, usually astringent.

Though well suited for use in wildlife plantings, there is nothing to commend this plant as an ornamental. However selections have played an important role in the development of the highly regarded interspecific sandcherry × plum hybrids.

Prunus × *cistena*
Purple-Leaved Sandcherry

Size	1.5 m.
Form	Upright spreading.
Texture	Medium.
Foliage	Deep-purple.
Flowers	White to pinkish, borne after the leaves.
Fruit	Dark purple, inconspicuous.
Bark	Deep purple; cambium, also purple.

Flowers and foliage are an attractive spring combination. This plant is not fully hardy. Tip killing is common in most locales, however dead portions of stem can be easily removed without spoiling the form of the plant.

Prunus × *cistena*

Prunus fruticosa
Mongolian Cherry

Size	1 m.
Form	Upright-spreading and closed to base.
Texture	Medium.
Foliage	Dark green, glossy leaves.
Flowers	White, in two- to four-flowered clusters.
Fruit	Bright red sour cherry, 1.5 cm, on individual peduncles.

This plant is hardy, but because of a strong suckering habit it can be used as an ornamental only where suckering is not a problem.

Prunus fruticosa

Prunus japonica
Chinese Bush Cherry

Size	0.5 m.
Form	Globular.
Texture	Medium.
Foliage	Dull green. Autumn foliage, yellow to orange.
Fruit	Dull red, tart, not prominent because it is borne beneath the foliage.

The size of the plant is its main ornamental characteristic. It does not have a strong infrastructure and therefore has little winter value. Autumn foliage is attractive.

Prunus japonica

Prunus maackii

Prunus maackii
Amur Cherry

Size	12 m.
Form	Upright-spreading, high- or low-headed.
Habit	Decurrent.
Canopy	Dense.
Texture	Coarse.
Foliage	Green, turning light yellow in the fall.
Flowers	Off-white, abundant.
Bark	Very attractive, golden or bronze-colored exfoliating bark.
Fruit	A small, black, astringent cherry.

This is a quick-growing tree with an outstanding winter habit. It is hardy in all parts of the region although subject to a form of low-temperature injury which occurs on cold nights when the bark tissues of the trunk contract more quickly than the wood they surround. This results in a long vertical split down the trunk. A more serious problem with this plant is its susceptibility to the shoe-string fungus *Armillaria meallia* though perhaps the splitting of the bark is a contributing factor. The fungus results in the early demise of the tree. When trees are infected, the bark changes to a deep copper color and the vigor of the tree soon begins to decline. Because of the fungus, the effective life of the tree is reduced to about 15 years. It is some consolation to know that this species grows quickly from seed.

Prunus maackii

Prunus nigra
Canada Plum

Prunus nigra

Size	5 m.
Form	Low-headed, ball-shaped.
Habit	Decurrent.
Canopy	Dense.
Texture	Coarse.
Foliage	Dull green.
Flowers	White, produced in abundance.
Bark	Gun-metal grey; branches with or without thorns.
Fruit	A plum of variable size and quality.

Canada plum is a good, small, spring-flowering tree for locations where the low-headed characteristic will not interfere with the activities of people. This species has been used in breeding programs leading to the development of better edible plums for the prairies.

Prunus × *nigrella* 'Muckle'
Muckle Plum

Size	3 m.
Form	Low-headed, upright-oval.
Habit	Decurrent.
Canopy	Closed.
Texture	Medium.
Foliage	Green.
Flowers	Bright pink, from bright red flower buds.

This is a spectacular small spring-flowering tree. It is an interspecific hybrid between the native Canada plum and the popular spring flowering shrub, the Russian almond.

Prunus × *nigrella* 'Muckle'

Prunus padus var. *commutata*
Mayday Tree, European Bird Cherry

Size	12 m.
Form	High-headed, upright-spreading.
Habit	Decurrent.
Canopy	Open or closed.
Texture	Coarse.
Foliage	Green.
Flowers	Fragrant, white, in loose pendulous racemes.
Fruit	Small, like chokecherry.

This tree has the following problems associated with it:

- A tendency to produce constantly forking branches which cause branch splitting
- It is subject to attack from two insect pests, the forest tent caterpillar *(Malacosoma spp.)* and an unidentified aphid
- The raisin-sized fruit tends to drop when ripe and is frequently picked up on shoes. Because of dropping fruit the tree should not be grown adjacent to decks and paved surfaces where people walk or congregate.

Prunus padus var. *commutata*

Prunus padus var. *commutata*

Woody Ornamentals

Prunus pensylvanica

Prunus pensylvanica
Pin Cherry

Size	4 m.
Form	Low-headed, upright-oval.
Habit	Decurrent.
Canopy	Open or closed.
Texture	Coarse.
Foliage	Green.
Flowers	Produced freely, small, white, showy.
Bark	Red, smooth, with horizontal lenticels.
Fruit	A small, bright red cherry, borne freely in small corymbs.

The pin cherry is to be found naturally on the sunny dry edges of clearings. The fall color of foliage is bright orange.

Cultivars of *Prunus pensylvanica*

'Jumping Pound' Small, weeping form of the species; less than 4 m.

'Mary Liss' A selection with fruit sizes up to three times that of the species.

'Stockton' A double-flowered form of the species but similar in most other respects.

Prunus pensylvanica 'Mary Liss'

Prunus × 'Prairie Almond'

Prunus × 'Prairie Almond'
Prairie Almond

Size	2 m.
Form	Globular to broad-oval, closed to base.
Texture	Medium.
Foliage	Green.
Flowers	Semi-double, pale pink with a purplish center.

Filaments of the flower are a deeper red-purple than those of the petals giving the effect of a darker eye. Flowers appear before the leaves.

Prunus pumila
Sandcherry

Size	0.3 m tall, 1-2 m in diameter.
Form	Low-spreading, leafy shrub suitable for underplanting or as groundcover.
Texture	Medium.
Foliage	Greyish green.
Flowers	White, borne in clusters of two to four.
Fruit	Almost globular, dark purple.

This can be a very useful, low plant for growing at the base of larger materials in a shrub grouping. It is not to be confused with the western sandcherry, *Prunus besseyi*, which is commonly a much taller shrub and not nearly as useful.

Prunus pumila

Prunus tenella
Russian Almond

Size	1 m.
Form	Medium-sized shrub; globular to broad-oval.
Texture	Fine.
Foliage	Green.
Flowers	Deep pink, showy.
Fruit	A small, bitter almond with a hairy pericarp.

The Russian almond has a tendency to sucker, so should be used in those places where the suckering habit is not objectionable.

Prunus tenella

Prunus tomentosa
Nanking Cherry

Size	2 m.
Form	Upright-spreading.
Texture	Medium.
Foliage	Deeply veined, dull green, tomentose.
Flowers	White, pinkish at petal fall.
Fruit	A bright red cherry to 1.5 cm, good for eating out of hand and for preserving.

One of the better large flowering shrubs, it does not color up in the fall.

Prunus tomentosa

Woody Ornamentals

Prunus triloba

Prunus triloba
Flowering Plum

Size	2–3 m.
Form	Upright-spreading.
Texture	Medium.
Foliage	Green, three-lobed.
Flowers	Single, white or pale pink.
Bark	Smooth, with horizontal lenticels.

Unfortunately this plant is not hardy in all parts of the region. In the north it is only crown hardy and because it blossoms only on wood of the previous season it can not be counted on as a flowering shrub. From the north-central part of the region south it can be relied upon fully. The insulating effect of snow can be an important factor in the protection of flower buds of this plant and many others that blossom on wood of the previous season.

Prunus triloba 'Multiplex'

Cultivars of *Prunus triloba*

'Multiplex' (double flowering plum) One of the more popular large flowering shrubs. It is similar in size and form to the species but has fully-double pink flowers produced in clusters along the stems before the leaves have formed. Buds are deep red and the flowers are shell pink.

Prunus virginiana 'Sharon'

Prunus virginiana
Chokecherry

Size	2-7 m.
Form	Upright-spreading.
Texture	Coarse.
Foliage	Green.
Flowers	White, in long, pendant racemes.
Fruit	Juicy, but astringent. Fruit color of the native western variety is typically black, but a range from yellow, red to dark purple exists.

This native plant is commonly found on dry slopes and at the sunny edge of the aspen forest. Some cultivars show a strong suckering tendency and may be grafted on a non-suckering rootstock to counter the problem.

Cultivars of *Prunus virginiana*

'Boughen's Chokeless' A selection with non-astringent fruit.

'Boughen's Yellow' A yellow-fruited type with larger than average sized fruit.

'Copper Shubert' A tree with an open canopy and similar in size to Shubert chokecherry. A tree with attractive copper red leaves. Fruit a dark copper red and less astringent than 'Shubert'.

'Mini-Shubert' A very compact, small tree with an upright oval habit. Interior branches are small and very upright. Mature leaf color similar to 'Shubert'.

'Sharon' A neat, low-headed tree with deep purple foliage. This plant is superior to 'Shubert' in that it is neater and requires less maintenance.

'Shubert' A popular ornamental tree to 7 m. The leaves of this plant emerge and size-up green but as they age they turn reddish purple. By midsummer all foliage will have turned color. Fruit is deep purple, moderately astringent and edible when cooked. This is a somewhat "weedy" tree in that it requires annual pruning to retain a good structural growth habit. It has a tendency to sucker on its own roots.

Prunus virginiana 'Shubert'

Pseudotsuga

Douglas-Fir

Two botanical varieties of Douglas-fir are indigenous to western North America, the coastal form *Pseudotsuga menziesii* and *Pseudotsuga menziesii* var. *glauca*, the form found growing in interior B.C. and as far east as Calgary. The foliage of the Douglas-fir is more like that of *Abies* than that of any of the other conifers; however, characteristics of the seed cone and of the terminal bud, which in this case is shiny brown and sharp-pointed, make positive identification easy.

Pseudotsuga menziesii var. *glauca*
Douglas-Fir

Size	20 m.
Form	Young trees are narrowly pyramidal; older trees are tall and straight-trunked, with a flat-topped crown.
Foliage	Needles, glossy blue-green above, white-lined beneath.
Bark	On young trees, smooth, grey with resin blisters; on older trees, thick and deeply-furrowed, reddish brown.
Seed Cones	Cylindrical, tapering toward the tips. A three-pronged bract protrudes from behind each cone scale on mature cones.

Young trees are neat and attractive; older trees are massive and picturesque.

Pseudotsuga menziesii var. *glauca*

Woody Ornamentals

Pyrus

Pear

The pears of commerce are not hardy in the region; however, those species of Eurasian origin are generally able to tolerate the severity of the western Canadian winter. Unfortunately the fruit produced by these is not of good quality. Several hybrids between the Ussurian pear and the commercial cultivar 'Bartlett' were introduced from the University of Saskatchewan but none can be considered as acceptable fruit plants. The cultivar 'Ure', an introduction from the Morden Research Station, is worthy of trial as a tree with edible fruit.

As ornamentals, the hardy pears are worth growing for their flowers and their autumn color. Flowers are perfect, white, about 4 cm in diameter and are produced early in the spring.

Pyrus ussuriensis

Pyrus ussuriensis

Pyrus ussuriensis
Ussurian Pear

Size	6 m.
Form	Low-headed, upright-oval.
Habit	Decurrent.
Canopy	Dense.
Texture	Coarse.
Foliage	Green.
Flowers	White, abundant, produced in clusters, very showy.
Fruit	Small, the size and shape of a crab apple; flesh is very gritty and hard.

Ussurian pear is a hardy plant, well adapted to most climates in the region, though shelter from wind is important in the south. It is not generally recommended for alkali soils. While this is an excellent flowering tree, it does not bloom fully until it is roughly 10 years old. Branches are well armed with large strong thorns. It is not a good plant for use on or adjacent to a lawn area because of its tendency to drop its fruit in late midsummer. The fruit is produced in great abundance; it breaks down very rapidly and can present significant problems when it gets picked up on shoes or when removal is called for prior to cutting the lawn.

Quercus

Oak

Oaks are deciduous trees with mostly upright-oval to oval heads. Leaves are short-petioled and deeply-lobed. Flowers are imperfect/monoecious; staminate flowers are produced in slender pendulous catkins; pistillate flowers are solitary or in two- to many-flowered spikes. The fruit is an acorn. Large trees are difficult to transplant because of long tap-roots and whorls of secondary roots. For this reason the use of containerized nurserystock is recommended when trees greater than 2 m are called for.

Quercus alba
White Oak

Size	20 m.
Form	Broad round-topped head.
Habit	Decurrent.
Canopy	Open.
Texture	Medium because of its lobed deeply-cleft leaves.
Leaves	Deeply-lobed leaves, bright green above, dull silver beneath. Rounded lobes defined by at least eight sinuses that reach almost to the yellowish green midrib. Leaf base narrowly wedge-shaped.
Flowers	Sessile, with the two forms often existing on the same branch.
Fruit	The acorn is ovoid with only one-quarter of the base enclosed in the shallow cup.

Quercus alba

A vigorous tree with attractive foliage.

Quercus borealis
Northern Red Oak

Size	10 m.
Form	Upright-oval head.
Canopy	Closed.
Texture	Coarse.
Foliage	Dark green, shiny above. Leaf-lobes sharp-pointed. Fall color is yellow to brown.
Fruit	A large acorn with only the basal 10% held in a very shallow cup.
Flowers	Very pendulous staminate catkins.

Quercus borealis

Northern red oak is not common, yet it is a good looking large tree. Mature specimens observed in the north-central parts of the region appear to be quite hardy.

Quercus coccinea

Scarlet Oak

Size	6 m.
Form	High-headed, upright-oval.
Habit	Decurrent.
Canopy	Open.
Texture	Coarse.
Foliage	Green, oblong, deeply-cleft, thin leaves with rounded sinuses defining seven, sometimes nine, lobes.
Fruit	Almost one-half the acorn enclosed in a hemispheric cup.

Quercus coccinea

This tree turns glowing scarlet in the fall. The color develops gradually branch by branch. It is rarely seen in the west, but trees growing in the city of Saskatoon seem to have no apparent difficulty. This plant certainly is worthy of trial.

Quercus macrocarpa

Bur Oak

Size	15-20 m.
Form	High-headed, upright-oval.
Habit	Decurrent, though frequently branching on younger trees tends to be horizontal.
Canopy	Closed.
Texture	Coarse.
Foliage	Green, obovate. Leaves are lobed with a large somewhat compound terminal lobe. The two lobes near the mid-point of the leaf have deep sinuses that reach almost to the mid-rib. but lobes closer to the the wedge-shaped base are relatively small.
Fruit	An acorn 2–2.5 cm long, more than half enclosed by a cup with a fringed border.

Quercus macrocarpa

The bur oak is a good tree over most of the region. It is said to be tolerant of fumes from automobile exhaust. Mature natural stands in south-eastern Manitoba are striking trees. Autumn color is antique yellow.

Quercus macrocarpa

Quercus mongolica
Mongolian Oak

Size	9 m.
Form	Upright-oval.
Habit	Decurrent.
Canopy	Open.
Texture	Coarse.
Foliage	Green with wavy margins; leaves tend to cluster at the ends of branches. Foliage turns color in the fall and tends to stay on the tree into the winter.
Fruit	Short-stalked, ellipsoidal acorn about 2 cm long, one-third within a thick cup.

This is a vigorous tree with a somewhat slender, though often irregular, habit.

Quercus mongolica

Rhamnus
Buckthorn

Buckthorns are deciduous shrubs, occasionally tree-like, sometimes thorny. They are drought-resistant. Plants are polygamous or dioecious, with small greenish yellow or white flowers, in axillary clusters, umbels or racemes. Flowers are not usually showy. Fruit is a berry-like drupe. Leaf veins are a distinguishing characteristic in that they tend to grow almost parallel to one another.

Rhamnus alnifolius
Alder-Leaved Buckthorn

Size	From 0.3 to 0.5 m.
Form	Neat, low spreading shrub, closed to base.
Texture	Medium.
Foliage	Glossy green, narrow, acute; 7.5 cm long with deeply-indented parallel veins.
Flowers	Greenish, single or in few-flowered umbels.
Fruit	Black, poisonous.

This plant is a compact, low-shrub which is good for massing and for low hedging. The autumn coloration is an outstanding feature. It is a clear, bright yellow.

Rhamnus alnifolius

Woody Ornamentals

Rhamnus frangula

Rhamnus frangula
Glossy Buckthorn

Size	3 m.
Form	Upright-spreading shrub.
Texture	Medium.
Foliage	Smooth, glossy green, attractive.
Bark	Stems are dark, flecked with grey lenticels.
Flowers	White in clusters of two to ten.
Fruit	Red, turning dark purple.

This is a neat, clean shrub. It is fully hardy within larger cities. It may tip-kill to some degree in rural areas and in the northern parts of the region. It is well suited to moist locations. The cultivar 'Columnaris' that is sold under the name "Tall Hedge" has been tried in several areas and has not proven to be hardy enough in the region.

Rhus

Sumac

The sumacs are small to large-sized deciduous shrubs, valued chiefly for their autumn coloration. In this region good autumn color cannot be expected from all species but the handsome, large, compound leaves of two of the larger ones are outstanding while still green. The plants are of easy culture. Some species have a tendency to produce suckers, a characteristic which makes them of some value as reclamation plants for roadsides and steep embankments in rural areas. Flowers are perfect, plants polygamous or dioecious. Fruit clusters are not showy, but the conical form of those produced by the two larger species is an interesting and distinctive feature.

Rhus glabra

Rhus glabra
Smooth Sumac

Size	3–4 m.
Form	Upright, leggy.
Texture	Coarse.
Foliage	Large pinnately-compound leaves. Dark, glossy green on upper surface, lighter beneath.
Flowers	Imperfect in stout panicles; male flowers yellow; female flowers green, turning red following pollination. Plants are dioecious or polygamous.
Fruit	Reddish; small, in narrowly conical, dense, hairy clusters.

This is a plant of easy culture. It is tolerant of atmospheric pollution. Large pinnately-compound leaves are uniquely ornamental. The plant has a tendency to sucker if the roots are disturbed and because of this it has some value for stabilizing slopes. Autumn coloration is seldom fully developed before leaves are killed by hard frost. A deep, somewhat sombre purple is typical of what might be expected in most parts of the region.

Cultivars of *Rhus glabra*

'Midi' A compact plant but not as compact as the cultivar "Mini".

'Mini' A neat compact globe to 1 m.

Rhus glabra

Rhus trilobata
Lemonade Sumac

Size	1 m.
Form	Mound-like, closed to base
Texture	Medium.
Foliage	Green, leaves trifoliate, aromatic when crushed.
Flowers	Yellow, in clusters, insignificant.
Fruit	A small, red globular drupe.

This attractive hardy small native shrub is particularly well suited to the hotter, drier parts of the region. It is a good plant for massing.

Rhus trilobata

Rhus typhina
Staghorn Sumac

Size	2–4 m.
Form	Upright, thicket forming, tall-shrub.
Texture	Coarse.
Foliage	Light green. Attractive large pinnately-compound leaves. Autumn color bright orange to red.
Flowers	Imperfect, greenish in dense terminal panicles. Plants are dioecious.
Bark	Wood of the current season is covered with dense, velvety brown hair.
Fruit	Reddish, hairy, in conical clusters.

Because of its open, straggly form and its suckering habit, it should be considered for use only in special situations. It prefers a sunny location and will do well on poor soil. Fall color is generally good.

Rhus typhina

Ribes

Currant, Gooseberry

These are deciduous shrubs, frequently with prickles. The leaves are alternate, usually palmately lobed. Plants are polygamous or dioecious, with flowers produced in few- or many-flowered racemes. The fruit is a juicy, many-seeded berry.

Ribes alpinum

Ribes alpinum
Alpine Currant

Size	1 m.
Form	Dense, ball-shaped, full to the base.
Texture	Medium.
Foliage	Dark green, lobed leaves.
Flowers	Greenish yellow, imperfect; borne in upright racemes. Plants are dioecious.

This is a good, dense, compact shrub for sun or shade. It is a good hedge plant and a good shrub for urban conditions because of its ability to withstand air pollution. It is often subject to powdery mildew in shaded sites.

Ribes oxycanthoides

Ribes oxycanthoides
Northern Gooseberry

Size	1 m.
Form	Slender, low plant with an abundance of thin wiry, curved stems.
Texture	Fine.
Foliage	Light green. Fall color soft oranges and reds.
Flowers	White, perfect.
Fruit	A reddish purple berry. Good for eating when ripe.
Bark	Branches are covered with sharp slender prickles.

This native plant is suitable against an appropriate background such as a low wall. It prefers moist sites.

Rosa

Rose, Shrub Rose

The shrub-rose is likely the oldest flower in cultivation and certainly the largest flowered shrub grown on the prairies. Many are not considered fully hardy in the region, but the more tender species and cultivars will generally survive when given winter protection.

Members of the genus are variable but hybridize easily. All those grown here are deciduous shrubs with upright stems, usually prickly, but sometimes unarmed. Leaves are compound, alternate, pinnate, and stipulate. Flowers are solitary or corymbose at the ends of short branchlets. The pistils of the flowers are numerous and are enclosed in an urn-shaped receptacle which becomes fleshy at maturity enclosing several bony achenes. These fleshy fruit-like structures are commonly known as "rose hips".

Rosa acicularis
Prickly Rose

Size	1 m.
Form	Upright-spreading.
Texture	Fine.
Foliage	Green, five to seven leaflets per leaf.
Flowers	Single, deep pink.
Bark	Covered with numerous straight, weak prickles.
Fruit	Pear-shaped "hip" with distinct neck.

The flower of this plant is the floral emblem of Alberta. It is not generally grown as an ornamental but native stands are sometimes encouraged. Insect pests associated with this species attack cultivated roses; therefore, the two should not be grown within insect-migrating distance of one another.

Rosa acicularis

Rosa arkansana
Sunshine Rose

Size	0.3 m.
Form	Low-spreading.
Texture	Medium.
Foliage	Green.
Flowers	Light pink, produced over a long season.

The wild plant is not grown in gardens but *R. arkansana* has three characteristics which have made it attractive to rose breeders: a dwarf habit of growth, a long season of bloom, and an ability to withstand drought. In the northern part of the region the *R. arkansana* hybrids show varying degrees of hardiness and all should be given winter protection.

Rosa arkansana

Hybrids of *Rosa arkansana*

× **'Adelaide Hoodless'** Plant height is 1 m; blossoms are borne in many-flowered clusters, semi-double to double with about 25 petals; color is medium to medium dark red. The plant blooms from early summer till frost. As a cut flower it is long-lasting if cut at the red-tip bud stage. Foliage is glossy green. Plants tend to sprawl with the weight of blossoms.

× **'Assiniboine'** Plant height is 1 m; blossoms are reddish purple with yellow stamens, ten petals per flower, produced singly or in corymbs. It blooms in mid-summer and occasionally throughout the summer.

Rosa arkansana × 'Assiniboine'

× **'Cuthbert Grant'** Plant height 0.5 -1m; blossoms are large, dark red, velvety with 15 firm petals. Flowers are borne in clusters of three to six on new growth in July and late summer. Winter-killing of tops can be expected on unprotected plants.

× **'Morden Amorette'** Plant height is 0.45 m; blossoms are 7–8 cm, red, semi-double with 25–30 petals; borne in small to large clusters. The plant is extremely floriferous and it blossoms continuously while growing conditions prevail. It kills to the snow line without protection.

× **'Morden Cardinette'** Plant height is 0.75 m; flowers are 8 cm with 25 cardinal-red petals. Flowers are borne singly or in clusters on strong stems. It is said to be more winter hardy than comparable roses.

Rosa arkansana × 'Cuthbert Grant'

× **'Morden Centennial'** Plant height is 1.5 m; flowers are 10 cm, fully double with 45 rose-bengal petals. Flowers are slightly fragrant, borne singly or in clusters of 15. It is a good shrub for garden decoration. It blooms on both old and new wood, so even when cut back heavily it can be relied upon.

× **'Morden Ruby'** Plant height is 1 m; flowers are 7–8 cm, fully double with 75 ruby-red petals. Flowers are borne in clusters and resemble roses of the floribunda class. It blooms freely in early summer and again in early autumn. Blossoms are long lasting. Plants are not winter hardy without protection.

Rosa blanda
Smooth Rose

Rosa blanda 'Betty Bland'

Size	1-2 m.
Form	Upright-spreading.
Texture	Fine.
Foliage	Green.
Flowers	Borne singly, light pink.
Bark	Smooth, shiny, red without prickles.

The wild plant is not usually grown in gardens but the species, which is native in northern Ontario and Manitoba, has had a part in the improvement of shrub roses for the prairies.

Hybrids of *Rosa blanda*

× **Betty Bland** A tall plant (to 2 m) and one that produces fully-double, two-tone pink blossoms over a long season.

× **Therese Bugnet** A multispecific hybrid with some *R. blanda* in its ancestry. It blooms from early summer till frost. Flowers are large, double and deep pink. Up to 12 buds are produced per shoot. Height is 2 m.

Rosa blanda × 'Therese Bugnet'

Rosa foetida
Austrian Brier Rose

Size	2 m.
Form	Upright-spreading.
Texture	Fine.
Foliage	Green, five to nine leaflets.
Flowers	Solitary or in clusters, deep yellow with an unpleasant odor early in the season.
Bark	Prickly with strong straight prickles.

This shrub is not often planted; its cultivars and hybrids are seen much more frequently. The yellow rose cultivars are susceptible to the disease black spot *(Diplocarpon rosae)* which can cause severe defoliation. Hips are red.

Rosa foetida 'Bicolor'

Cultivars and Hybrids of *Rosa foetida*

× **'Bicolor'** (Austrian copper rose) A spectacular plant. Flowers are coppery-red outside, yellow inside.

× **'Harisonii'** (Harison's yellow rose) A widely planted double-yellow rose to 2 m. It has little fragrance. Fruit is black. This is an interspecific hybrid with *R. spinosissima*.

'Persiana' (Persian yellow rose) Flowers fully double. Blossoms deeper yellow and have more petals than Harison's Yellow.

Rosa foetida 'Harisonii'

Rosa gallica × 'Grandiflora'

Rosa gallica
French Rose

Size	1–2 m.
Form	Upright.
Texture	Coarse.
Foliage	Deep green.
Flowers	Brilliant red, single.

The particular combination of flower and foliage color is striking. The flower color lacks the purple undertones that so many roses possess. Flowers are produced over a short season. The species appears in the pedigree of the tender hybrid tea rose.

Cultivars of *Rosa gallica*

× **'Grandiflora'** Large single flowers to 8 cm, deep red, fragrant.

Rosa rubrifolia

Rosa rubrifolia
Red-Leaf Rose

Size	2 m.
Form	Upright; arching branches.
Texture	Medium.
Foliage	Reddish purple.
Flowers	Light pink, single

The foliage color combines well with other deciduous plants. The plant is susceptible to *Agrobacterium radiobacter* which is responsible for crown gall, a disease, the symptoms of which show up on the crown of the plant just below the soil surface.

Rosa rugosa 'Hansa'

Rosa rugosa
Rugosa Rose

Size	1 m.
Form	Upright-spreading.
Texture	Coarse.
Foliage	Light to medium green, wrinkled, often glossy.
Flowers	Solitary or in few-flowered clusters.
Bark	Grey, with stout prickles.
Fruit	Sub-globose, smooth and brick-red, 2.5 cm in diameter.

Flowers are produced over a long season, beginning in early summer.

Hybrids of *Rosa rugosa*

× **'Blanc Double de Coubert'** White semi-double flowers; blush tinted in bud. Fragrant.

× **F.J. Grootendorst** A hybrid between *R. rugosa* and *R. polyantha*. A 0.6–1 m plant with clusters of small, deep pink flowers with frilled edges This plant blooms in mid-summer.

× **Hansa** A ball-shaped shrub with double, purplish red fragrant flowers. Blooms are recurrent. Insect galls on stems are fairly common.

× **Martin Frobisher** A tall shrub to 2 m with fragrant flowers from early summer until frost. Flowers are two-toned, soft pink, being somewhat darker at the base of the petals in the center of the flower. This plant has red stems with very few prickles.

Rosa rugosa 'Hansa'

Rosa spinosissima
Altai Scots Rose

Size	2-3 m.
Form	Upright, leggy plant.
Texture	Fine.
Foliage	Green, leaflets 5-11.
Flowers	Single, creamy white, 6-9 cm across, very fragrant.
Bark	Very prickly.
Fruit	Hip sub-globose, shiny black.

This is one of the earliest roses to bloom. It is not a recurrent bloomer but there is often a second flush of bloom in late summer. It suckers freely and its form is not attractive.

Rosa spinosissima

Hybrids of *Rosa spinosissima*

× **'Yellow Altai'** A hybrid between *R. spinosissima* and × 'Harisonii'. It resembles *R. spinosissima* but has yellow flowers.

var. *altaica* Tall, leggy shrub with white "single" flowers; suckering habit.

Rosa spinosissima var. *altaica*

Woody Ornamentals

Salix

Willow

In spite of the lack of a good hardy weeping-willow, there are many willows of both shrub and tree type suited to western conditions. Some are grown for their leaf color and tree form, others for bark color and still others (pussy willows) for the male catkins that are produced early in the spring.

Willows possess alternate, usually lance-shaped leaves, which emerge from buds covered by a single bud-scale. The flowers of all species are imperfect with plants being dioecious. "Pussy willows" are produced by the male tree. Female trees produce a longer catkin and a fruit that splits when mature to release minute seeds that are surrounded by tufts of silky hair. In this respect the willow is like its relative, the poplar, but the fluff produced is not nearly as noticeable nor as offensive.

Young branches of the shrubby willows are usually long, straight, unbranched, flexible and highly colored. For these reasons, they are well suited to basket weaving and related pursuits.

Willows are moisture-loving and do not stand up well to very dry conditions. Because they make vigorous growth when conditions are right, they would be expected to make a good type of tree for shelterbelt purposes; however, because they have such a strong moisture dependency, their value as shelterbelt trees is limited unless they can be irrigated. Willows are easily propagated from dormant cuttings because they possess pre-formed root initials. In most cases it is only necessary to stick dormant branches into moist soil in order to obtain new plants.

Salix acutifolia

Salix acutifolia
Sharp-Leaf Willow

Size	6 m.
Form	Upright-spreading, frequently multi-stemmed.
Habit	Decurrent.
Canopy	Dense, closed.
Texture	Medium.
Foliage	Green, carried to the ground.

Leaves are subject to attack from a gall-forming insect. Basal shoots are thin and untidy, giving a shrubby appearance to the base of the tree. It is a good shelterbelt tree where sufficient moisture is available.

Salix alba 'Sericea'

Salix alba
White Willow

Size	20 m.
Form	Broad-oval, low-headed.
Habit	Decurrent.
Canopy	Dense.
Texture	Medium.
Foliage	Billowing masses of silvery leaves.

A good tree but not as common as the three cultivars listed. Like all willows this species prefers moist to wet sites.

Cultivars of *Salix alba*

'Chermesina' (red-barked white willow) An upright spreading plant that is generally grown as a shrub. Stems are cut to the ground each spring to promote vigorous new osiers which develop good red color for winter effect.

'Sericea' (Siberian white willow) An upright oval tree to 12 m with good silvery foliage which can be seen from some distance. Some difficulty has been experienced with getting this tree established. However it is fully hardy. It does not always respond well to fall planting.

'Vitellina' (golden willow) Quite similar to the species but with bright golden bark on the newest growth. This is a good tree for large landscapes where it can be recognized from a great distance in early spring before leaf-out. It is also good for irrigated shelterbelts where it is frequently seen in multi-stemmed form.

Salix alba 'Vitellina'

Salix brachycarpa 'Blue Fox'
Blue Fox Willow

Size	1 m.
Form	Globular plant closed to base.
Texture	Medium.
Foliage	Blue-grey.

This small, shrubby willow has some value because of its foliage color.

Salix brachycarpa 'Blue Fox'

Salix daphnoides
Daphne Willow

Size	6 m.
Form	Upright oval, female trees narrower and more columnar than male trees.
Habit	Decurrent.
Canopy	Dense.
Texture	Medium.
Foliage	Dark lustrous green above, bluish beneath. Stipules at the base of the petiole are large and semi-cordate.
Flowers	The 5-cm male catkins, produced before the leaves, are dense and very showy; they are good for bouquets.
Bark	Branches and shoots are long, violet-purple overlaid with a white bloom.

Very ornamental during winter and spring.

Salix daphnoides

Salix discolor

Pussy Willow

Size	4 m.
Form	Large oval shrub or small tree.
Texture	Medium.
Leaves	Elliptic-oblong, glaucous on the undersides.
Flowers	Typical pussy willow catkins on male plants.

These plants are common to wet spots and are highly valued by admirers of pussy willows. In the early spring the woolly catkins come out long before the leaves. It is occasionally planted.

Salix discolor

Salix exigua

Coyote Willow

Size	4–5 m.
Form	Upright-spreading.
Texture	Fine.
Foliage	Linear to linear-lanceolate, silvery.
Bark	Silvery grey.

The silvery foliage color of this species is outstanding. The plant has one major problem in that it suckers very freely. Because of its grassy foliage, however, the plant lends itself to planting in containers to simulate bamboo.

Salix exigua

Salix interior

Sandbar Willow

Size	3 m.
Form	Upright-spreading.
Texture	Very fine.
Foliage	Green.

It is adapted to wet sites. It has value only as a plant for reclamation purposes.

Cultivars of *Salix interior*

'Fangstad' (Fangstad weeping willow) A small tree with excellent weeping habit. Unfortunately, it has weak ground support and is of little value on its own roots.

Salix interior

Salix lanata
Woolly Willow

Size	0.5 m.
Form	Procumbent.
Texture	Medium.
Foliage	Silvery grey, downy leaves.
Flowers	Catkins erect, yellowish grey.

The woolly willow is a useful shrub for base plantings and rock gardens. It prefers full sun. It may not be hardy in northern areas.

Salix lanata

Salix pentandra
Laurel-Leaf Willow

Size	12–15 m.
Form	Low-headed, broad-oval.
Habit	Decurrent.
Canopy	Dense.
Texture	Coarse.
Foliage	Dark, glossy green.

These trees have massive heads, short, stout trunks, and root systems that are not all that stable in a structural sense. They should not be grown in unusually wet places because of this instability. The species is hardy throughout the region and can be easily started from either softwood or hardwood cuttings. Excellent climbing tree for youngsters because of its low headed character and deeply furrowed bark.

Salix pentandra

Salix purpurea 'Gracilis'
Purple-Osier Willow

Size	1 m.
Form	Mound-like.
Texture	Fine.
Foliage	Linear, blue-green.
Bark	Purplish.

Stems are straight and non-branching and fan out at ground level from a small crown. It is a useful low shrub for mass effect. Tip killing can be expected in some areas.

Salix purpurea 'Gracilis'

Sambucus

Elder

Elders are shrubs that generally have a loose-open habit of growth. Occasionally the species *S. racemosa* is seen in tree form. Branches of all species are stout and pithy. Leaves are compound and opposite. Flowers are small and white in terminal clusters. The fruit is a small berry-like drupe and is displayed in conspicuous clusters.

Sambucus canadensis

Sambucus canadensis
American Elder

Size	2 m.
Form	Ball-shaped.
Texture	Medium.
Foliage	Dark green
Flowers	White, in flat, terminal clusters.
Fruit	Purple to black.

Fruit of this plant is frequently used for jellies and wines.

Sambucus nigra 'Aurea'

Sambucus nigra 'Aurea'
Golden Elder

Size	2 m.
Form	Upright-spreading.
Texture	Coarse.
Foliage	Bright yellow when growing in full sun.
Flowers	White, in large, flat-topped heads.
Fruit	Black.

A plant of easy culture and at one time very common. It has gradually been replaced by more dependable and more attractive cultivars of *S. racemosa* even though it is now considered to be fully hardy in most areas.

Sambucus racemosa
Red Elder

Size	3 m.
Form	Upright-spreading with arching branches.
Texture	Coarse.
Foliage	Green, compound, with deeply toothed margins.
Flowers	Small, in large, cream-colored clusters.
Fruit	Small, but red clusters are quite showy.

The red elder is a large, coarse, leggy shrub that can be grown almost anywhere space will allow. It is a native species and sometimes found in the boreal mixed-wood forest. It is sometimes used as a small single-stemmed tree. It is susceptible to a debilitating fungus disease which has been fairly common in recent years. Because the disease affects the vascular system, one of the symptoms is wilting of the foliage.

Sambucus racemosa

Cultivars of *Sambucus racemosa*

'Goldenlocks' A dwarf 75 cm, ball-shaped shrub with golden, finely cut, leaved foliage.

'Plumosa Aurea ' A form with dull golden foliage. Leaves deeply incised.

'Redman' A heavy fruiting selection of the species with deeply incised, frequently dissected leaves.

'Sutherland Golden' A bright golden tall shrub with finely dissected leaves.

Sambucus racemosa 'Sutherland Golden'

Sedum
Stonecrop

The low ground-hugging "succulents" belonging to this genus are useful groundcover plants.

Sedum kamtchaticum
Orange Stonecrop

Size	0.04–0.08 m.
Form	Vegetative mat with an interesting texture.
Texture	Medium.
Foliage	Tight matted clusters of light-green leaves.
Flowers	Small, yellow blossoms, borne above the foliage in midsummer.

This little plant lends itself to use as a groundcover on low, sloping south-facing banks. When planted on level ground it seldom has the opportunity to fully display its interesting characteristics.

Sedum kamtchaticum

Sedum spurium 'Dragon's Blood'

Sedum spurium 'Dragon's Blood'
Dragon's Blood Two-Row Stonecrop

Size	0.05 – 0.1 m.
Form	Thick, flat tangle of short, snake-like stems that spread horizontally to cover the ground.
Texture	Medium.
Foliage	Clusters of deep red succulent leaves.
Flowers	Deep red, borne above the foliage in late midsummer.

This is the most colorful of the groundcovers. While the foliage provides an interesting tapestry during the growing season it is in fact exceeded by the additional color at the time of blossoming. Dead-heading the old flower heads at the end of the season may be a problem for some gardeners.

Shepherdia

Buffaloberry

These native shrubs or shrubby trees in the west have some value as ornamentals. Flowers are small and imperfect, and appear early. Plants are dioecious, with fruit on female trees small, juicy and edible; varying from the typical red to yellow. Leaves are opposite and silvery.

Shepherdia argentea

Shepherdia argentea
Silver Buffaloberry

Size	3 m.
Form	Roughly ball-shaped, but with a very irregular branching habit. Secondary branches often at 90 degree angles.
Texture	Fine.
Foliage	Strap-shaped, silvery on both sides.
Flowers	Yellow, in small short pedicelled clusters at nodes.
Bark	Silvery with stout thorns up to 5 cm long.
Fruit	Deep red, in tight clusters along stems.

This is a shrub or small tree with rugged, twisting, thorny branches. It is well adapted to dryland situations. With skilled pruning, this plant can be easily transformed into an interesting accent tree for special landscape effect. Users should be aware of the fact that it can produce suckers at unpredictable distances from the adult plant.

Cultivars of *Shepherdia argentea*
'Goldeye' Similar to the type but with yellow fruit.

Shepherdia argentea

Shepherdia canadensis
Russet Buffaloberry

Size	1 m.
Form	Ball-shaped, closed to the base.
Texture	Medium.
Foliage	Dark, dull green above, silvery with brown scales on underside.
Flowers	Yellow, but small and not very conspicuous.
Fruit	Variable, red to orange.

The russet buffaloberry has a neat, dense habit of growth but tends to become loose and open in shade. It prefers well-drained soils when moisture is available. It is a native plant that is deserving of wider use.

Shepherdia canadensis

Sorbaria

False-Spirea

These tall, suckering, thicket-forming shrubs have compound leaves that resemble those of mountain-ash. They are well adapted to shady places, although they also do well in full sun. The leaf habit gives them a fine-textured appearance. Flowers are produced in fluffy white terminal panicles 5–30 cm long. They are not highly valued as landscape plants but if this is not a problem and the requirement is for a shrub with mid-summer flowers, then the false-spirea may have some value.

Woody Ornamentals

Sorbaria sorbifolia
Ural False-Spirea

Size	2 m.
Form	Upright, thicket-forming plant.
Texture	Medium.
Foliage	Dull green.
Flowers	White, turning light brown as they fade; panicles 15 cm long.

This plant is highly susceptible to spider mite infestation when drought stressed. It will grow almost anywhere.

Sorbaria sorbifolia

Sorbus

Mountain-Ash

These deciduous trees are valued highly for their dark green foliage, their showy clusters of fragrant, white flowers, and attractive scarlet fruit. Leaves are alternate and pinnately compound. Flowers appear in late spring and are borne in compound, terminal corymbs. The fruit is a small pome which varies in taste from bland to bitter depending on the species. Bark varies from dark brown to orange-brown, and is usually smooth although in at least one species it becomes roughened and scale-like with age.

Sorbus americana
American Mountain-Ash

Size	4–6 m.
Form	Upright oval, low-headed.
Habit	Decurrent.
Canopy	Dense.
Texture	Medium.
Foliage	Dark green, good orange fall color.
Flowers	White in large clusters; fragrance heavy.
Bark	Dark brown.
Fruit	Scarlet, in large terminal clusters.

This is a good tree because of its many ornamental characteristics. It is hardier and much more reliable than the European species. Fruit is attractive to winter birds like the waxwing and grosbeak. Winter buds are a distinguishing feature. In this species they are black, sharp-pointed and sticky.

Sorbus americana

Sorbus aucuparia
European Mountain-Ash, Rowan Tree

Size	6 m.
Form	Globular, low-headed.
Habit	Decurrent.
Canopy	Dense.
Texture	Medium.
Foliage	Dark green. Autumn color is flame red in late fall.
Flowers	White.
Fruit	Red.

Because it is somewhat slow to ripen its wood at the end of the growing season, it is sometimes subject to winter injury. It is more susceptible to sunscald than are the North American species. Winter buds are covered with dense grey hair and are not sticky.

Sorbus aucuparia

Cultivars of *Sorbus aucuparia*

'Fastigiata' This cultivar is a tight, columnar form with stiff erect branches. The fruit is scarlet in dense bunches. It is a good tree where space is limiting. Since it is subject to sunscald, the trunk should be protected from the sun during the winter.

Sorbus aucuparia 'Fastigiata'

Sorbus scopulina
Western Mountain-Ash

Size	5 m.
Form	Globular, usually multi-stemmed.
Habit	Decurrent.
Canopy	Dense.
Texture	Medium.
Foliage	Medium-dark green. Develops bright orange to red fall colors; showy.
Flowers	White.
Fruit	Orangy red.

This is a large, round, shrubby native tree, usually with several good strong stems, large fruit clusters and good soft autumn coloration occurring in early autumn. It is suited to moist soils and sunny exposures.

Sorbus scopulina

Spiraea

Spirea

Spireas are a group of small to medium-sized deciduous shrubs, widely grown because of their decorative flowers. Flower color ranges from deep pink to white. A few species are native but these are not often used in the cultivated landscape. Most cultivated types have been introduced.

Spiraea × arguta

Spiraea × arguta
Garland Spirea

Size	1 m.
Form	Upright-spreading.
Texture	Fine.
Foliage	Greyish green.
Flowers	White.

The garland spirea is free-flowering and showy in early summer. Blossoms are produced on arching branches of the previous season's wood. This plant has little to commend it once it is through blossoming. It has a tendency to become leggy with many upright stems and requires a good deal of annual thinning.

Spiraea × billiardii

Spiraea × billiardii
Billiard Spirea

Size	1 m.
Form	Upright plant with a suckering habit.
Texture	Medium.
Foliage	Green.
Flowers	Pink flowers in dense terminal panicles.

The billiard spirea blooms in mid-summer. It is well suited for woodland edge situations, but the very upright, thicket-forming habit does not lend itself to other use.

Spiraea × *bumalda*
Dwarf Pink Spirea

Size	0.6 m.
Form	Low-spreading.
Texture	Medium.
Foliage	Green with red overtones.
Flowers	Produced on wood of the current season, deep pink in flat-topped clusters.

Blooms over most of the summer. Like the hybrid tea rose, it has a "cut and come again" habit and in order to perpetuate continuous blossoming old flower heads must be pruned out (not sheared) when they start to deteriorate. A useful small shrub for massing in the garden in sunny locations.

Spiraea × *bumalda* 'Anthony Waterer'

Cultivars of *Spiraea* × *bumalda*

'Anthony Waterer' Deep pink blossoms; height 0.6 m. The hardiest cultivar in the north.

'Crispa' Small plant with deeply serrated, twisted leaves. Blossoms are brighter, more intense pink, and leaves a brighter green than those of the species. Although not commonly available it is highly recommended.

'Dart's Red' Very tight, mound-forming plant with deep red flowers. Foliage color varies, although red predominates.

'Froebelii' Light red blossoms; height 0.5 m.

'Gold Flame' Dull pink blossoms, but with an interesting foliage which varies from reddish through orange to yellow as the season progresses. More compact than either 'Froebelii' or 'Anthony Waterer' and useful for massing.

Spiraea × *bumalda* 'Gold Flame'

Spiraea japonica
Japanese Spirea

Size	0.75 m.
Form	Upright-globular.
Texture	Medium.
Foliage	Green, coarsely-toothed.
Flowers	Pink in large flattened heads.

The species is less well known than its cultivars and not commonly seen.

Cultivars of *Spiraea japonica*

'Alpina' A neat compact, horizontal plant to 0.30 m. Very floriferous with pink flowers carried 5 cm above the foliage.

'Goldmound' A very dense dwarf, compact ball-shaped plant with light yellow leaves and small, light pink flowers.

'Little Princess' A low compact-mound with rose-colored flowers. Well maintained plants will bloom all season long. Autumn foliage color red and quite showy.

Spiraea japonica 'Goldmound'

Spiraea media sericea

Oriental Spirea

Size	1 m.
Form	Compact mound.
Texture	Medium.
Foliage	Green.
Flowers	White in late June.

This medium-sized shrub has a neat rounded form, closed to its base. Because of its growth habit, it is very useful in mass plantings in the larger landscape.

Spiraea media sericea

Spiraea trilobata

Three-Lobed Spirea

Size	1 m.
Form	Ball-shaped, closed to base.
Texture	Fine.
Foliage	Dark green.
Flowers	White.

This is the finest of the white spireas and possibly the whitest. It has a very neat habit of growth, something that is not common among the white spireas. Even the old inflorescenses are retained in an attractive way and therefore do not require dead-heading.

Spiraea trilobata

Spiraea × vanhouttei

Bridalwreath Spirea

Size	1.5 m.
Form	Upright-spreading with arching branches.
Texture	Medium.
Foliage	Green.
Flowers	White, in arching sprays in early summer.

This hybrid spirea has been around for a long time and has outgrown its usefulness. Like many of the white spireas it tends to produce an abundance of canes which give the plant a very crowded and unattractive appearance.

Spiraea × vanhouttei

Syringa

Lilac

The lilacs are tall, vigorous, mostly coarse-textured shrubs, sometimes trees, with large, dark to medium-green leaves. Flowers are perfect, produced in large terminal or lateral panicles from buds formed on wood of the previous season. Flowers are usually very fragrant, sometimes double, and appear during spring and early summer. There is considerable variety in flower color, and because each of the various types bloom over a long period, lilacs can offer much more to the home owner than most other groups of woody ornamentals. Because of their large size, however, plants can be out of scale with the small property if used too freely. In recent years, the Lilac in some parts of the region was attacked by a leaf-mining insect *(Gracillaria syringella)*. The insect attacks only the thinner-leaved species typified by *S. vulgaris*. Fortunately the systemic insecticide, dimethoate, when used as a soil drench early in the growing season, has given a good measure of control. Leaf-miner infestations appear to be cyclic and at the time of publication (1995) we appear to be at the low point of the cycle.

Syringa × chinensis
Rouen Lilac

Size	2 m.
Form	Upright-spreading.
Texture	Medium.
Foliage	Green, much narrower than most.
Flowers	Purple, in large loose panicles.

The relationship between lilac and privet is evident in the leaves of the Rouen lilac. The form is quite compact, and the plant is non-suckering and is useful for hedging.

Syringa × chinensis

Syringa × hyacinthiflora
Hyacinth-Flowered Lilac

Size	3 m.
Form	Upright-spreading.
Texture	Coarse.
Foliage	Green, thin, much like those of the parents, *S. vulgaris* and *S. oblata* var. *dilitata*.
Flowers	Clusters large, showy, fragrant.

This interspecific hybrid is one of the best of the tall, early flowering lilacs. It can always be relied upon to produce large trusses of fragrant flowers. All cultivars of *S. × hyacinthiflora* are fully hardy and highly recommended.

Syringa × hyacinthiflora 'Minnehaha'

Cultivars of *Syringa* × *hyacinthiflora*

Syringa × *hyacinthiflora* 'Sister Justina'

'Assessippi' Broad-petaled, single-flowers, argyll purple.

'Minnehaha' Single flowers, bishop's purple.

'Pocahontas' Single flowers, dark purple.

'Sister Justina' Pure white, double flowers.

'Swarthmore' Very hardy and reliable. Flowers lilac purple and fully double.

Syringa × *josiflexa*
Josiflexa Lilac

Syringa × *josiflexa* 'Guinivere'

Size	3 m.
Form	Narrowly-upright, leggy plant.
Texture	Coarse.
Foliage	Green, leaves thick.
Flowers	Nodding panicles of lilac purple.

The main feature of this plant is its nodding flower panicles. It is a useful tall plant for the back of the shrub border.

Cultivars of *Syringa* × *josiflexa*

'Bellicent' Pink flowers and graceful growth.

'Guinevere' With short broad panicles of deep argyll purple.

'Lynette' Pink, in loose-flowered panicles. Foliage mottled.

Syringa meyeri
Meyer Lilac

Syringa meyeri

Size	2 m.
Form	Ball-shaped, closed to base.
Texture	Medium.
Foliage	Green with wavy margins.
Flowers	Small, compact violet-pink panicles, in late spring.

This compact plant is fully hardy and useful in the shrub border. Flowers are carried over the whole plant from top to bottom.

Syringa oblata var. *dilitata*
Korean Early Lilac

Size	3 m.
Form	Upright-spreading.
Texture	Coarse.
Foliage	Green, occasionally turning deep wine red in autumn.
Flowers	Clusters large, showy, fragrant.

This plant is rarely seen in the region and is not fully hardy. However, it has been the parent of the hybrid species *S. × hyacinthiflora*, the best of the early lilacs.

Syringa oblata var. *dilitata*

Syringa × persica
Persian Lilac

Size	2 m.
Form	Ball-shaped.
Texture	Medium.
Foliage	Light green, small by lilac standards and sometimes lobed.
Flowers	Pale lilac in broad panicles.

The Persian lilac is not widely grown but still is a useful shrub for the small home grounds. It may not be fully hardy in some of the outlying areas.

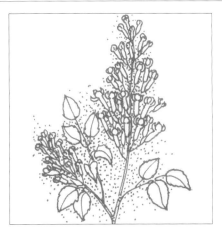

Syringa × persica

Syringa × prestoniae
Preston Lilac

Size	3 m.
Form	Ball-shaped.
Texture	Coarse.
Foliage	Green.
Flowers	Single, color variable, pink to purple.

Preston lilac cultivars are popular, late-blooming (early-summer) lilacs. They are very much superior to both the late lilac species *S. villosa* and *S. josikaea*. They are of easy culture and have no particular preferences.

Syringa × prestoniae 'Minuet'

Cultivars of *Syringa × prestoniae*

'Coral' Compact plant, with clear pink flowers.

'Donald Wyman' Reddish purple. Outstanding branching habit.

'Isabella' Large dense plant, flower trusses, pink.

'James MacFarlane' Dense flower clusters, deep pink.

'Jessica' Open flower trusses, deep argyll purple.

'Minuet' Dense, dwarf-compact plant with light purple flowers.

'Miss Canada' Clear, bright China-rose flowers borne freely in narrow spike-like panicles.

Woody Ornamentals

Syringa reticulata

Syringa reticulata
Japanese Tree Lilac

Size	5 m.
Form	Upright-spreading. Both single and multi-stemmed trees are found.
Habit	Decurrent.
Canopy	Dense.
Texture	Coarse.
Foliage	Green.
Flowers	Cream-colored, in large trusses.
Bark	Bark is an attractive, deep brown, and smooth, like that of cherry, with horizontal lenticels.

This tree blossoms in early to mid-summer after all the other lilacs have finished blooming. It is very showy. The Japanese tree lilac is not hardy in the north but the variety *S. reticulara* var. *amurensis* is, and it is strongly recommended as a substitute in northerly parts of the region. It is a large shrub with large trusses of blossoms that are similar to those of the species.

Syringa × swegiflexa

Syringa × swegiflexa
Chengtu Lilac

Size	3 m.
Form	Upright-spreading; leggy.
Texture	Coarse.
Foliage	Green.
Flowers	Deep red, fading to pink.

The nodding flower trusses are the outstanding feature of this hybrid species.

Cultivars of *Syringa × swegiflexa*

'Fountain' A *S. swegiflexa* × *S. reflexa* hybrid backcrossed to *S. reflexa*. The extra dose of the latter has resulted in a plant that produces an abundance of nodding, pale pink flower trusses in late June.

Syringa velutina
Manchurian Lilac

Size	3 m.
Form	Upright-spreading.
Texture	Coarse.
Foliage	Outstanding, velvety, deep green foliage.
Flowers	Trusses large, loose and lacy, tinged light purple.

The combination of flowers and foliage makes this large shrub worthy of much wider use. It is fully hardy with no particular problems.

Syringa velutina

Syringa villosa
Late Lilac

Size	3 m.
Form	Ball shaped.
Texture	Coarse, thick.
Foliage	Light, dull green.
Flowers	Rosy-lilac.

This is not a good landscape plant because it lacks good ornamental characteristics. Flower trusses are small and sparse. It is well suited, however, to the dry exposed conditions that are to be encountered in the south, where it has been used in shelterbelts.

Syringa villosa

Syringa vulgaris
Common Lilac

Size	3 m.
Form	Upright-spreading with suckering habit.
Texture	Coarse.
Foliage	Dark green.
Flowers	Purple.

The species has a bad reputation because of its strong suckering habit. However, it does have fragrant blossoms which are freely produced. It is a good shrub for situations where suckering is not a problem.

Like some other lilac species, this one is the parent of a race of popular ornamentals known as the French hybrids or French grafted lilacs. Many of these produce double flowers. This is also one of the thinner-leaved species that in recent years has become subject to attack by leaf-mining insects.

Syringa vulgaris 'Charles Joly'

Woody Ornamentals

Hybrids of *Syringa vulgaris*

Syringa vulgaris 'Charm'

× **'Congo'** Flowers single, magenta in large, compact panicles.

× **'Charm'** Single, pink.

× **'Charles Joly'** Large panicles of double reddish purple flowers.

× **'Edith Cavell'** Double-flowered, white.

× **'General Pershing'** Double-flowered, deep purple.

× **'Katherine Havemeyer'** Double-flowered, cobalt blue, flushed mauve.

× **'Ludwig Spaeth'** Flowers single, purple.

× **'Madame Lemoine'** Double-flowered, creamy white, narrow panicles.

× **'Montaigne'** Double-flowered, pinkish mauve.

× **'Mrs. E. Willmott'** Double-flowered, white.

× **'Mrs. Edward Harding'** Double-flowered, bright purplish red.

Tamarix

Tamarisk

These shrubs have very slender branches and very fine leaves. Small branchlets are deciduous and fall with the leaves. Leaves are scale-like, often sheathing, like juniper. Flowers are small, short-petioled or sessile, in dense racemes, that are usually concentrated in terminal panicles. The texture of these shrubs is extremely fine and feathery. The flowers, which are pink or pinkish, are also fine-textured and appear like puffs of pink smoke on the airy green foliage. Tamarisk cannot be considered fully hardy.

Tamarix pentandra

Tamarix pentandra
Amur Tamarisk

Size	2–3 m.
Form	Upright-spreading.
Texture	Very fine.
Foliage	Light green.
Flowers	Pink in terminal panicles.

This plant cannot be considered fully hardy, but is well worth trying. Flowers are produced in late summer on wood of the current season. Bare root plants do not transplant readily, however, it has been found that plants can be established from hardwood cuttings if the unrooted cutting is stuck in moist soil in the place where the plant is to grow.

Taxus

Yew

The yews, hardy enough for the prairie provinces, are dwarf conifers with attractive dark green, usually glossy, flat needles. Plants are dioecious. For a conifer the fruit is unusual in that it is a fleshy berry-like "cone", about 1 cm in diameter and coral to deep pink in color. Most parts of the Japanese yew are poisonous if eaten, however, although the seed is poisonous the red aril enclosing it is not.

Yews are not frequently grown in the region, but it is not uncommon to see small trees or large spreading shrubs happily growing in some well-sheltered location. Experience has shown that the foliage of most cultivars is subject to winter desiccation on exposed sites. Doubtless they prefer a moister climate than what is available in the prairie provinces. Nevertheless, the yew is a plant that is well worth trying in suitable locations.

Taxus cuspidata 'Nana'
Dwarf Japanese Yew

Size	1 m.
Form	Globular.
Texture	Fine.
Flowers	Imperfect.
Foliage	Dark green above, lighter beneath; mid-rib slightly raised.

Seed Cones On female plants only; deep coral pink.

This is a good, hardy, dark green conifer for sheltered places where the foliage can be protected from winter desiccation.

Taxus cuspidata 'Nana'

Thuja

Arborvitae, White Cedar

The cultivars hardy in the region are very dense, slow-growing plants, which ultimately form columnar, pyramidal or globular shrubs. Because of their distinct and unusual habit of growth, the columnar and pyramidal forms are commonly used as vertical accents.

In eastern Canada, where the white cedar is native, it is frequently used as a hedge plant, but in the west this is hardly practical due to the maintenance required in a dry climate.

The white cedar is shallow-rooted and so requires supplementary water under most western conditions. Leaves are flat, plate-like and overlapping (like the scales of a fish), with the lateral leaves nearly covering the facial ones. Flowers are imperfect/monoecious. Branches grow in either vertical or horizontal planes.

Like junipers, white cedars are subject to injury from both the early spring sun and the red spider mite.

Woody Ornamentals

Thuja occidentalis 'Brandon'

Thuja occidentalis 'Little Giant'

Thuja occidentalis 'Robusta'

Thuja occidentalis 'Skybound'

Thuja occidentalis
American Arborvitae

Size	Most within the range 0.5–6 m.
Form	Columnar, but pyramidal and globular forms are common among the cultivars.
Texture	Fine.
Foliage	Green.
Flowers	Imperfect/monoecious
Seed Cones	Scaly, small, egg-shaped, light brown when mature.

All plants in this species are moisture-loving and require supplementary moisture in most parts of the region. Because they are susceptible to desiccation from winter conditions, direct exposure to sun and wind should be avoided.

Cultivars of *Thuja occidentalis*

'Brandon' and **'Carman Columnar'** Good columnar forms to 6 m derived from western-grown material. The foliage is deep green.

'Columnaris' One of the better columnar cedars; it is superior to both 'Brandon' and 'Carman Columnar'.

'Danica' A slow-growing, small, globular plant with yellowish green foliage held in vertical sprays. It has one of the best globular forms but requires a well-sheltered site.

'Globosa' This is a compact-globe shrub to 1 m with a tendency to spread laterally as it gets older.

'Hetz Midget' An extremely dwarf globular plant to 0.30 m with attractive foliage which is laid down in almost vertical layers.

'Holmstrup' A narrow-conical form to 6 m with better foliage than either 'Brandon' or 'Carman Columnar'. Leaf masses have a layered look and are carried in vertical planes.

'Little Gem' Small, dense, globular form to 60 cm with touselled, glossy green foliage.

'Little Giant' A small, dense, ovoid form to 0.9 m.

'Robusta' Sometimes referred to as Wares Siberian cedar, it has a broad pyramidal form to 3 m, is dull, dark green, turning purplish during the winter. It is not suitable for the home grounds because of its large size.

'Skybound' Narrowly-columnar with dark green foliage.

'Smaragd' A neat upright-columnar shrub to 6 m with flat upright sprays of bright green foliage. It is said to be more subject to winter sunscald than some others of this type. This plant has a distinctive tapered tip and is a very formal-looking plant.

'Techny' A loose, broad-columnar plant to 6 m, this plant can make a good, tall, untrimmed hedge or tree for the shrub border but is far too large for the small home grounds. Its form is not particularly interesting.

'Woodwardii' A very large, globular shrub to 2 m, with dense, light green foliage.

Tilia

Basswood, Linden

These are large, flowering trees with a dense, compact habit of growth. Leaf size is generally large, but it varies considerably between species. The shape of the tree is pyramidal but may become more open with age. Leaves are alternate with slender petioles. Leaf bases are heart-shaped or oblique; margins are usually serrate but are occasionally coarsely-toothed. Terminal buds are lacking. Flowers, which are borne in pendulous clusters, are yellowish, conspicuous and fragrant. Flower stems are joined to strap-shaped bracts. The fruit is a hard nutlet, covered with a thin green husk.

Because of the denseness of the foliage, the main trunk seldom gets enough light to support good tangential growth. This usually results in a central leader that rapidly diminishes in girth in the upper third of the tree.

These trees are attacked by the linden mite, a tiny insect which produces small, light green galls that project straight out from the surface of the leaves. The galls are from 4–6 mm long and about 2 mm in diameter. Since the insect living within the gall cannot be hit with a conventional insecticide, use of the systemic insecticide, dimethoate, is the only practical means of control. Two applications, a year apart, are required for complete control.

Since damage caused by the mites is mainly cosmetic, treatment should only be undertaken on highly valuable trees in prominent situations.

Tilia americana
Basswood

Size	15 m.
Form	Low-headed, pyramidal; open when fully mature.
Habit	Excurrent.
Canopy	Dense.
Texture	Coarse.
Foliage	Dark green, large with obliquely cordate base.
Flowers	Yellow, fragrant, in pendulous cymes.

This is a useful large tree for most situations and one adapted to most parts of the region, with the possible exception of the northernmost areas.

Tilia americana

Tilia cordata
Little-Leaf Linden

Size	10–12 m.
Form	Low-headed, pyramidal.
Habit	Excurrent.
Canopy	Dense.
Texture	Medium.
Foliage	Green, round with a cordate base.
Flowers	White to yellow, fragrant, in pendulous cymes.
Bark	Smooth, reddish brown.

This is a very popular tree because of its neatness, its flowering habit and its clear-yellow autumn coloration.

Tilia cordata

Tilia cordata

Cultivars of *Tilia cordata*

'Greenspire' This is a neat, small, compact tree that is considered the best of the lindens for street planting.

'Morden' Considered to be one of the hardiest cultivars, this tree is similar to the species in form and density but may be slightly smaller.

Tilia × flavescens

Tilia × flavescens
Dropmore Linden

Size	10 m.
Form	Pyramidal.
Habit	Excurrent.
Canopy	Dense.
Texture	Coarse.
Foliage	Dark green.
Flowers	Perfect; yellow in drooping cymes with pedicels attached to unique strap-shaped bracts.
Bark	Smooth, furrowed with age.

This is a hybrid between *T. americana* and *T. cordata* and one that is well adapted to western conditions. It is resistant to the linden mite. Following bloom the tree will produce a super-abundance of small woody fruit which will drop in late summer.

Cultivars of *Tilia × flavescens*

'Wascana' A selected seedling with stronger, wide-angled branches and a faster growth rate.

Tilia mongolica

Tilia mongolica
Mongolian Linden

Size	5 m.
Form	Low-headed, broad-pyramid.
Habit	Excurrent.
Canopy	Dense.
Texture	Coarse.
Foliage	Green, leaves coarsely toothed, sometimes lobed.
Flowers	Yellow.
Bark	Rough, in plates.
Fruit	Pear-shaped, the size of a pea.

Foliage jaggedly toothed, bright yellow in early September.

Ulmus

Elm

The elms are trees that have been widely used for street planting and shade. The leaves are short-stalked, usually doubly saw-toothed. Leaf bases are asymmetrical. Flowers are perfect, appearing in spring before the leaves, in small dense clusters or in racemes. The fruit, which ripens a few weeks after flowering, is flat, one-seeded, and surrounded by a broad membranous wing. Elms grow readily from seed planted immediately after it has matured. Growth of the seedlings is rapid, and one species, *U. pumila*, will make about 60 cm of growth per year once established.

Ulmus americana
American Elm

Size	30 m.
Form	High-headed, upright-spreading.
Habit	Decurrent.
Canopy	Open.
Texture	Medium.
Foliage	Green.

These trees have narrow-angled, weak crotches. They have been good street trees, but are too large for small properties. Dutch elm disease, which has already devastated the American elm in eastern Canada, has been found in native stands in Saskatchewan; hence use of this tree is not encouraged. No fully resistant selections are known.

Cultivars of *Ulmus americana*

'Beaverlodge' A well formed American elm tree, recommended for northern areas.

'Brandon' A very neat clone with a dense, compact head due to many ascending branches. Each branch forms a very narrow, angled, weak crotch with the main trunk. This could very well show up as a major weakness as the trees mature.

Ulmus americana

Ulmus americana 'Brandon'

Ulmus davidana var. *Japonica*

Ulmus davidiana var. *Japonica*
Japanese Elm

Size	12 m.
Form	High-headed, upright-spreading.
Habit	Decurrent.
Canopy	Open.
Texture	Medium.
Foliage	Green.

Form and hardiness of this tree is similar to that of the American elm although it is a much smaller tree. It has been successfully hybridized with American elm in an attempt to produce a tree resistant to Dutch elm disease. The hybrid 'Jacan' was considered resistant but has recently gone down to the disease.

Ulmus pumila

Ulmus pumila
Manchurian Elm

Size	10 m.
Form	High-headed, upright-spreading.
Habit	Decurrent.
Canopy	Open.
Texture	Fine.
Foliage	Green.

This weedy tree has many bad habits. It possesses weak crotches. It produces an abundance of fruit which is dropped in the spring over a three-week period. It has deciduous branchlets which fall freely in response to the slightest breeze, and it is also very subject to slime flux, a bacterial disorder which causes foul-smelling bacterial ooze to emerge from interior wood at crotches and wound sites. The only quality that might attract interest is the fact that it grows very quickly and it can be used as a hedge plant.

Viburnum

Viburnum, Bush-Cranberry

These are highly ornamental deciduous spring-flowering shrubs. Some species make attractive small trees. Leaves are coarse-textured, lobed or entire. Inflorescence is white or pinkish, showy in umbel-like or paniculately-compound cymes. In some species the showy inflorescence is due to prominent clusters of sterile flowers. In these species fruit is produced from fertile flowers which form a visually insignificant part of the inflorescence. The fruit is generally showy.

Viburnum lantana
Wayfaring Tree

Size	1.5 m.
Form	Ball-shaped.
Texture	Very coarse.
Foliage	Grey-green.
Flowers	White, showy.
Fruit	Immature fruit is bright scarlet and very showy; unfortunately the fruit turns black when it matures.

This is a good ornamental due to its showy fruit and an often extended blossoming period. It does well in both sun and shade but takes on a much neater form in full sun. Winter buds are unique in that they are naked.

Viburnum lantana

Viburnum lentago
Nannyberry

Size	3-5 m.
Form	Upright-spreading and leggy, though compact forms exist.
Texture	Coarse.
Foliage	Glossy green. Brilliant orange and purplish red in autumn.
Flowers	White, showy.
Fruit	Black.

The flowering qualities of this plant are special, as is the autumn coloration. The plant also has good form. It has a tendency to sucker if the roots are disturbed. It does well in both sun and shade. Because the leaves droop from the petioles, they tend to appear slightly wilted. This is often the first plant to turn color in fall.

Viburnum lentago

Viburnum opulus
Guelder-Rose

Size	2 m.
Form	Upright-spreading.
Texture	Coarse.
Foliage	Green, three-lobed leaves.
Flowers	White.
Fruit	Scarlet.

This plant is similar to, but not as hardy as, the native high bush-cranberry. It prefers a moist, shady environment.

Viburnum opulus

Woody Ornamentals

Cultivars of *Viburnum opulus*

'Compactum' A dense compact shrub with a free flowering and fruiting habit.

'Nanum' A juvenile plant that does not produce flowers and fruit. It does well in shade. Autumn foliage is a dull purple. It is prone to insect infestation.

'Sterile' This form is known as "Snowball" because of the white sterile flowers which are gathered into large conspicuous globular heads. The plant is not fully hardy, likely due to a lack of root hardiness. Because the flowers are sterile, it cannot produce fruit.

'Xanthocarpum' A large shrub to 3 m with attractive golden yellow fruit.

Viburnum opulus 'Nanum'

Viburnum rafinesquianum
Arrowwood

Size	2 m.
Form	Upright-oval, dense habit.
Texture	Coarse.
Foliage	Dark green. Fall color good; dull wine red.
Flowers	White.
Fruit	Shiny, black.

The arrowwood is a striking ornamental of easy culture though rarely seen in cultivation. It is native to eastern Manitoba.

Viburnum rafinesquianum

Viburnum sargentii
Sargent's Viburnum

Size	To 5 m.
Form	Upright-spreading.
Texture	Coarse.
Foliage	Green. Autumn color good, bright orange to red.
Flowers	White.
Fruit	Red.

This Asiatic plant looks like a larger version of the American high bush-cranberry. It is very showy in flower and fruit and has spectacular autumn foliage color. It is superior to *V. trilobum* in the way that it displays its fruit.

Viburnum sargentii

Viburnum trilobum
High Bush-Cranberry

Size	3 m.
Form	Upright-spreading.
Texture	Coarse.
Foliage	Green, three-lobed. Autumn color good, bright orange to red.
Flowers	Inconspicuous, perfect, borne in umbel-like cymes and surrounded by conspicuous sterile florets.
Fruit	A juicy red, edible drupe.

This is a leggy shrub or small flowering tree that thrives in shady wet sites and yet one that will tolerate sites that are sunny and much drier. An excellent ornamental because its effects are expressed at three different times during the year: in late spring when it blooms, in the summer when it is in fruit, and again in the fall when the foliage colors up.

Cultivars of *Viburnum trilobum*

'Andrews' A compact form with good fruiting characteristics. A notable improvement over the species.

'Compactum' A shrub to 0.75 m with a tendency to produce straight, strong non-branching stems. This is a non-flowering clone. The plant colors up nicely in the fall but displays a stiffness during the winter because of its non-branching habit. However, when used in mass it makes a very respectable winter statement. The stem color is yellow.

'Garry Pink' A shrub with all the characteristics of the species except flower color. In this case the flowers are a very light shell pink and fully the size of a 25-cent piece.

Viburnum trilobum

Viburnum trilobum 'Compactum'

Viburnum trilobum 'Garry Pink'

Vinca

Periwinkle, Barvenok

The species of *Vinca* that are hardy in the three western provinces are both groundcovers. They are prostrate plants and spread by stolons. The leaves are ovate-lanceolate, opposite, and dark green on the upper surface. Flowers are five-petaled, tubular, flat on top, sky blue in color, and appear in spring and early summer. Both species discussed require a cool root-run and a shady location. Vinca was brought to the region by early Ukrainian settlers and is still locally known by the name "Barvenok."

Vinca herbacea

Vinca herbacea
Deciduous Periwinkle

Size	0.8 m.
Form	Prostrate with long, thread-like stolons that root readily at the nodes.
Texture	Fine.
Foliage	Lustrous green, deciduous.
Flowers	Sky blue, produced in abundance over a long season.

The deciduous periwinkle does well in sun and semi-shade. It is best used by itself because of its tendency to produce long, thread-like stolons that may not appeal to the busy gardener if they spread over other things. Still, it is a worthwhile plant if the gardener is prepared to reduce stolon activity.

Vinca minor 'Bowles'

Vinca minor 'Bowles'
Bowles Periwinkle

Size	0.1 m.
Form	Ground-hugging plant.
Texture	Medium.
Foliage	Glossy, dark green. Leaves evergreen.
Flowers	Conspicuous, attractive sky blue, in spring.

This is a good groundcover for shady areas. Even though it spreads by above-ground stems (stolons), it still prefers a deep friable soil containing plenty of organic matter. Snow cover during the winter is important for survival. In order to compete with perennial weeds during establishment, it must be planted not further than 12 cm apart.

Vitis

Grape

One of the earliest plants cultivated by humans but only occasionally is the grape seen growing in the prairie provinces; nevertheless, one native species can be found growing in Manitoba. It is not generally grown for the fruit, but more frequently as an ornamental vine for covering arbors and walls. The grape plant is polygamo-dioecious, that is, it will have both perfect and imperfect flowers on the same plant and yet the imperfect flowers will always be on separate plants. The grape climbs using tendrils to hold the vines to any adjacent vertical support.

Vitis riparia
Riverbank Grape

Size	6 m.
Form	A vigorous climber.
Texture	Coarse.
Flowers	Produced in late spring, flowers are small and inconspicuous but staminate ones are fragrant.
Foliage	Light green, lustrous, large, three-lobed.
Fruit	Small, blue with a light blue blush.

This very vigorous climber has been used as a plant for covering arbors. Because of its vigor, it does require a certain amount of pruning to keep it under control. The two cultivars listed are less vigorous and easier to manage than the species. Fruit is astringent, and of little value fresh, but has been used in making jams and jellies.

Recommended cultivars of *Vitis riparia*

× **'Beta'** A hybrid between *V. riparia* and the native grape of the eastern seaboard. The fruit appears much like that of 'Concord', only smaller.

'Valiant' A new cultivar from South Dakota State University that is reported to be earlier maturing and hardier than 'Beta'.

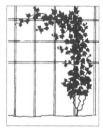

Vitis riparia

Vitis riparia

Reference Charts

Coniferous Shrubs

Plant Name	Size				Form								Foliage					
	Less than 0.3 m	0.3 – 0.6 m	1 – 2 m	2 – 6 m	Decumbent	Prostrate	Pyramidal	Columnar	Mound-like	Globe	Low Spreading	Tortuous	Deep Green	Yellow Green	Golden	Blue-Green	Grey-Green	Feathery
American Arborvitae																		
Brandon				•				•					•					
Carman Columnar				•				•					•					
Columnaris				•				•					•					
Danica		•								•				•				
Globosa		•								•			•					
Hetz Midget	•									•				•				
Holmstrup				•				•					•					
Little Gem		•								•			•					
Little Giant			•								•		•					
Robusta				•			•						•					
Skybound				•				•								•		
Smaragd				•				•								•		
Techny				•					•				•					
Woodwardii			•							•			•					
Dwarf Fir																		
Dwarf Balsam Fir		•				•										•		
Juniper																		
Common Ground Juniper		•									•		•					
Depressa Aurea		•				•									•			
Repanda	•					•							•					
Horizontal Juniper																		
Andorra Juniper	•										•					•		
Blue Chip	•					•										•		
Blue Prince	•															•		
Blue Rug Juniper	•										•					•		
Compact Andorra Juniper	•				•								•					
Dunvegan Blue	•					•										•		
Hughes		•									•					•		
Prince of Wales	•					•					•		•					
Turquoise Spreader	•					•										•		
Wapiti	•			•							•							
Waukegan Juniper	•				•						•						•	

Coniferous Shrubs

Seed Cones	Dry Conditions	Moist Conditions	Full Sun	Partial Shade	Full Shade	Sheltered	Unsheltered	Salt	Hardiness (?)	Negative Characteristics	Limitations	Page Number	Plant Name
													American Arborvitae
		•	•	•	•							160	Brandon
		•	•	•	•							160	Carman Columnar
		•	•	•	•							160	Columnaris
		•	•	•	•							160	Danica
		•	•	•	•							160	Globosa
		•	•	•	•							160	Hetz Midget
		•	•	•	•							160	Holmstrup
		•	•	•	•							160	Little Gem
			•	•	•							160	Little Giant
		•	•	•	•							160	Robusta
		•	•	•	•							160	Skybound
		•	•	•	•							160	Smaragd
			•	•	•							160	Techny
		•	•	•	•							160	Woodwardii
													Dwarf Fir
		•		•		•						33	Dwarf Balsam Fir
													Juniper
			•	•		•						84	Common Ground Juniper
			•	•		•						84	Depressa Aurea
				•		•						84	Repanda
												85	Horizontal Juniper
			•			•						85	Andorra Juniper
			•			•						85	Blue Chip
			•			•						85	Blue Prince
			•			•						85	Blue Rug Juniper
			•			•						85	Compact Andorra Juniper
			•			•						85	Dunvegan Blue
			•			•						85	Hughes
			•			•						85	Prince of Wales
			•			•						85	Turquoise Spreader
			•			•						85	Wapiti
			•			•						85	Waukegan Juniper

Woody Ornamentals

Coniferous Shrubs

Plant Name	Less than 0.3 m	0.3 – 0.6 m	1 – 2 m	2 – 6 m	Decumbent	Prostrate	Pyramidal	Columnar	Mound-like	Globe	Low Spreading	Tortuous	Deep Green	Yellow Green	Golden	Blue-Green	Grey-Green	Feathery
Jap-garden Juniper	•					•								•				
Pfitzer Juniper		•									•		•					•
Armstrongii		•									•		•					•
Aurea		•									•				•			•
Compacta		•									•		•					•
Glauca		•									•						•	•
Golden Pfitzer		•									•				•			
Mint Julep		•									•					•		
Rocky Mountain Juniper				•			•									•	•	
Blue Heaven				•			•									•		
Grey Gleam				•			•										•	
Grizzly Bear				•			•										•	
McFarland				•				•									•	
Medora				•			•	•								•		
Moffettii				•			•									•		
Wichita Blue				•			•										•	
Winter Blue		•									•					•		•
Savin Juniper			•								•		•					
Arcadia		•									•		•					
Broadmoor	•					•										•		
Blue Danube		•									•					•		
Calgary Carpet	•					•									•			
Hoar Frost	•					•									•			
Skandia	•										•						•	
Tamariscifolia		•							•				•					
Pine																		
Bristlecone Pine				•									•					
Mugo Pine, Swiss Mountain Pine			•						•	•	•		•					
Compacta			•							•			•					
Teeny		•							•				•					
Compact Scots Pine			•				•									•		
Sentinel Pine			•					•								•		

Coniferous Shrubs

Seed Cones	Dry Conditions	Moist Conditions	Full Sun	Partial Shade	Full Shade	Sheltered	Unsheltered	Salt	Hardiness (?)	Negative Characteristics	Limitations	Page Number	Plant Name
			•			•						86	Jap-garden Juniper
			•			•						86	Pfitzer Juniper
			•			•						86	Armstrongii
			•			•			•	•		86	Aurea
			•			•						86	Compacta
			•			•						86	Glauca
			•			•						86	Golden Pfitzer
			•			•			•			86	Mint Julep
			•			•						88	Rocky Mountain Juniper
			•			•						88	Blue Heaven
			•			•						88	Grey Gleam
			•			•						88	Grizzly Bear
			•			•				•		88	McFarland
			•			•						88	Medora
			•			•						88	Moffettii
			•			•						88	Wichita Blue
			•			•						88	Winter Blue
			•			•						87	Savin Juniper
			•			•						87	Arcadia
			•			•						87	Broadmoor
			•			•						87	Blue Danube
			•			•						87	Calgary Carpet
			•			•						87	Hoar Frost
			•			•						87	Skandia
			•			•						87	Tamariscifolia
													Pine
			•			•						108	Bristlecone Pine
			•			•						110	Mugo Pine, Swiss Mountain Pine
			•			•						110	Compacta
			•			•						110	Teeny
			•			•						112	Compact Scots Pine
			•			•						112	Sentinel Pine

Woody Ornamentals

Coniferous Shrubs

Plant Name	Size				Form								Foliage					
	Less than 0.3 m	0.3 – 0.6 m	1 – 2 m	2 – 6 m	Decumbent	Prostrate	Pyramidal	Columnar	Mound-like	Globe	Low Spreading	Tortuous	Deep Green	Yellow Green	Golden	Blue-Green	Grey-Green	Feathery
Dwarf Spruce																		
Black Hills Spruce				•			•						•					
Compact Norway Spruce			•				•						•					
Dwarf Alberta Spruce			•				•						•					
Dwarf Serbian Spruce			•				•									•		
Globe Colorado Spruce			•							•						•		
Dwarf Norway Spruce		•								•			•					
Nest Spruce	•										•		•					
Procumbent Colorado Spruce	•						•				•	•				•		
Yew																		
Dwarf Japanese Yew			•							•			•					

Coniferous Shrubs

Seed Cones	Dry Conditions	Moist Conditions	Full Sun	Partial Shade	Full Shade	Sheltered	Unsheltered	Salt	Hardiness (?)	Negative Characteristics	Limitations	Page Number	Plant Name
													Dwarf Spruce
				•		•						105	Black Hills Spruce
				•		•						105	Compact Norway Spruce
				•		•			•			106	Dwarf Alberta Spruce
				•		•						106	Dwarf Serbian Spruce
				•		•						107	Globe Colorado Spruce
				•		•						105	Dwarf Norway Spruce
				•		•						105	Nest Spruce
			•			•						107	Procumbent Colorado Spruce
													Yew
				•		•						159	Dwarf Japanese Yew

Woody Ornamentals

Coniferous Trees

Plant Name	Size				Form			Habit		Density		Foliage		
	1–3 m	5–10 m	10–15 m	15–20 m	Pyramidal	Columnar	Irregular	Excurrent	Decurrent	Dense	Open	Green	Blue-Green	Silvery Green
Douglas-Fir														
Douglas-Fir				•	•			•		•		•		
Fir														
Balsam Fir			•		•			•		•		•		
Siberian Fir			•		•			•		•		•		
Subalpine Fir			•		•			•		•		•		
White Fir, Colorado Fir		•			•			•		•				•
Larch, Tamarack														
Siberian Larch			•		•			•			•	•		
Tamarack			•		•			•			•	•		
Pine														
Austrian Pine				•	•			•		•		•		
Eastern White Pine				•	•			•			•		•	
Jack Pine		•					•		•		•	•		
Limber Pine		•			•			•			•	•		
Lodgepole Pine			•		•			•			•	•		
Red Pine, Norway Pine			•		•			•		•		•		
Scots Pine			•				•		•		•	•		
Swiss Stone Pine			•			•		•		•		•		
Western Yellow Pine, Ponderosa Pine				•		•		•		•		•		
var. *scopulorum*				•		•		•		•		•		
White Bark Pine			•		•			•			•	•		
Spruce														
Colorado Spruce				•	•			•		•		•	•	
Hoopsii			•		•			•		•			•	
Koster			•		•			•		•			•	
Moorheimii			•		•			•		•			•	
Montgomery	•				•			•		•			•	
Morden Blue			•		•			•		•			•	
Engelman Spruce				•	•			•		•		•		
Norway Spruce				•	•			•		•		•		
Serbian Spruce			•		•			•		•			•	
White Spruce				•	•			•		•		•		
Black Hills Spruce			•		•			•		•		•		

Coniferous Trees

Bark	Seed Cones	Dry Conditions	Moist Conditions	Full Sun	Partial Shade	Full Shade	Sheltered	Unsheltered	Salt Tolerant	Hardiness (?)	Negative Characteristics	Limitations	Page Number	
														Douglas-Fir
		•	•										127	Douglas-Fir
														Fir
				•	•	•				•			33	Balsam Fir
				•	•	•				•			34	Siberian Fir
				•	•	•				•			34	Subalpine Fir
			•			•							34	White Fir, Colorado Fir
														Larch, Tamarack
			•										89	Siberian Larch
		•	•										89	Tamarack
														Pine
	•			•									111	Austrian Pine
	•			•	•							•	112	Eastern White Pine
				•				•					109	Jack Pine
				•				•					110	Limber Pine
				•				•					109	Lodgepole Pine
				•				•					111	Red Pine, Norway Pine
•				•				•					112	Scots Pine
•		•		•			•						109	Swiss Stone Pine
				•				•					111	Western Yellow Pine, Ponderosa Pine
	•	•		•			•						111	var. *scopulorum*
•				•				•					108	White Bark Pine
														Spruce
				•				•					107	Colorado Spruce
				•				•					107	Hoopsii
				•				•					107	Koster
				•				•					107	Moorheimii
				•				•					107	Montgomery
				•				•					107	Morden Blue
		•	•					•					105	Engelman Spruce
			•					•					105	Norway Spruce
					•	•							106	Serbian Spruce
			•					•					106	White Spruce
			•					•					106	Black Hills Spruce

Special Features column group spans: Bark, Seed Cones, Dry Conditions, Moist Conditions, Full Sun, Partial Shade, Full Shade, Sheltered, Unsheltered, Salt Tolerant, Hardiness (?), Negative Characteristics, Limitations.

Woody Ornamentals

Deciduous Shrubs

Plant Name	Size				Form							Texture			Foliage					
	Less than 30 cm	0.3–1 m	1–2 m	2–5 m	Low-Spreading	Mound-like	Globe	Columnar	Upright Spreading	Leggy	Closed to Base	Fine	Medium	Coarse	Green	Variegated	Grey-Green	Purple, Red	Silver	Yellow
Barberry																				
Korean Barberry			•				•				•		•		•					
Poiret's Barberry			•				•				•		•		•					
Sheridan Red Barberry			•				•				•		•					•		
Broom																				
Golden Broom		•				•						•			•					
Buckthorn																				
Alder-Leaved Buckthorn		•			•								•	•						
Glossy Buckthorn			•						•				•	•						
Buffaloberry																				
Russet Buffaloberry		•					•				•		•		•					
Silver Buffaloberry			•							•			•						•	
Goldeye			•						•				•						•	
Burningbush, Spindletree																				
Aldenham Spindletree			•				•				•		•		•					
Dwarf Narrow-Leaved Burningbush	•						•					•			•					
Turkestan Narrow-Leaved Burningbush			•				•					•			•					
Dwarf-Winged Burningbush	•					•					•		•		•					
Maack's Spindletree				•					•	•			•		•					
Spindletree			•						•	•			•		•					
Wartybark Burningbush		•					•				•		•		•					
Winged Burningbush			•						•	•			•		•					
Wintercreeper		•							•				•		•					
Caragana																				
Common Caragana			•						•	•		•			•					
Lorbergii		•					•					•			•					
Pendula		•					•					•			•					
Plume			•				•					•			•					
Sutherland				•					•	•		•			•					
Tidy		•							•	•		•			•					
Walker		•					•					•			•					
Pygmy Caragana		•					•				•	•			•					
Russian Caragana			•						•	•		•			•					
Globosa		•					•					•			•					
Shagspine Caragana			•							•		•			•					

Deciduous Shrubs

White	Violet, Purple	Red, Pink	Yellow	Fragrant	Ornamental Bark	Ornamental Fruit	Dry Conditions	Wet Conditions	Full Sun	Partial Shade	Sheltered Site	Salt Tolerant	Tender	Negative Characteristics	Limitations	Autumn Color	Winter Effect	Page Number	Plant Name
																			Barberry
			•			•			•									47	Korean Barberry
			•			•			•							•		47	Poiret's Barberry
			•						•					•				47	Sheridan Red Barberry
																			Broom
			•						•									67	Golden Broom
																			Buckthorn
							•	•				•			•			131	Alder-Leaved Buckthorn
•						•			•	•								132	Glossy Buckthorn
																			Buffaloberry
			•			•			•	•								147	Russet Buffaloberry
			•			•	•		•			•						146	Silver Buffaloberry
			•			•	•		•			•						147	Goldeye
																			Burningbush, Spindletree
						•			•								•	71	Aldenham Spindletree
						•			•							•		72	Dwarf Narrow-Leaved Burningbush
						•			•							•		73	Turkestan Narrow-Leaved Burningbush
						•			•		•					•		71	Dwarf-Winged Burningbush
		•			•	•			•							•		72	Maacks Spindletree
						•			•							•		71	Spindletree
						•			•							•		73	Wartybark Burningbush
					•	•			•	•						•		71	Winged Burningbush
•						•				•			•					72	Wintercreeper
																			Caragana
			•				•		•			•						52	Common Caragana
			•				•		•			•						52	Lorbergii
							•		•			•						52	Pendula
							•					•						52	Plume
			•				•		•			•						52	Sutherland
			•				•		•			•						52	Tidy
							•		•			•						52	Walker
			•				•		•			•						53	Pygmy Caragana
			•				•		•			•						53	Russian Caragana
							•		•									53	Globosa
•							•		•			•						53	Shagspine Caragana

Woody Ornamentals

Deciduous Shrubs

Plant Name	Size				Form							Texture			Foliage					
	Less than 30 cm	0.3 – 1 m	1–2 m	2–5 m	Low-Spreading	Mound-like	Globe	Columnar	Upright Spreading	Leggy	Closed to Base	Fine	Medium	Coarse	Green	Variegated	Grey-Green	Purple, Red	Silver	Yellow
Cinquefoil																				
Shrubby Cinquefoil		•					•				•	•			•					
Abbotswood		•				•					•	•			•					
Coronation Triumph		•					•				•	•			•					
Forestii		•				•					•	•			•					
Goldfinger		•					•				•	•			•					
Jackmanii		•							•		•	•			•					
Katherine Dykes		•				•					•	•			•					
Parvifolia		•				•					•	•			•					
Red Ace		•					•				•	•			•					
Snow Bird		•					•				•	•			•					
Tangerine		•					•				•	•			•					
Yellow Bird		•					•				•	•			•					
Clematis																				
Fragrant Ground Clematis		•			•								•		•					
Cotoneaster																				
Brickberry Cotoneaster			•						•		•		•				•			
Creeping Cotoneaster	•				•								•		•					
European Cotoneaster		•					•				•		•		•					
Flowering Cotoneaster			•						•		•		•		•					
Hedge Cotoneaster			•					•			•		•		•					
Nan Shan Cotoneaster	•						•						•		•					
Roundleaf Cotoneaster	•				•									•	•					
Sungari Rockspray Cotoneaster			•				•						•				•			
Currant, Gooseberry																				
Alpine Currant		•					•						•		•					
Northern Gooseberry		•			•							•			•					
Daphne, Garlandflower																				
February Daphne		•					•					•		•	•					
Rose Daphne	•				•							•			•					

Deciduous Shrubs

White	Violet, Purple	Red, Pink	Yellow	Fragrant	Ornamental Bark	Ornamental Fruit	Dry Conditions	Wet Conditions	Full Sun	Partial Shade	Sheltered Site	Salt Tolerant	Tender	Negative Characteristics	Limitations	Autumn Color	Winter Effect	Page Number	Plant Name
																			Cinquefoil
			•						•									118	Shrubby Cinquefoil
•									•									118	Abbotswood
			•						•									118	Coronation Triumph
			•						•									118	Forestii
			•						•									118	Goldfinger
			•						•									118	Jackmanii
			•						•									119	Katherine Dykes
			•						•									119	Parvifolia
		•							•		•		•					119	Red Ace
•									•									119	Snow Bird
			•						•				•					119	Tangerine
			•						•									119	Yellow Bird
																			Clematis
•									•				•					57	Ground Clematis
																			Cotoneaster
						•			•									64	Brickberry Cotoneaster
						•			•	•						•		62	Creeping Cotoneaster
•						•			•									63	European Cotoneaster
•									•									64	Flowering Cotoneaster
						•			•							•		63	Hedge Cotoneaster
						•			•									62	Nan Shan Cotoneaster
						•			•	•	•			•				64	Roundleaf Cotoneaster
•						•			•									63	Sungari Rockspray Cotoneaster
																			Currant, Gooseberry
									•									134	Alpine Currant
							•	•							•			134	Northern Gooseberry
																			Daphne, Garlandflower
	•					•					•			•				68	February Daphne
		•		•					•	•		•						68	Rose Daphne

Woody Ornamentals

Deciduous Shrubs

Plant Name	Size				Form							Texture			Foliage					
	Less than 30 cm	0.3 – 1 m	1–2 m	2–5 m	Low-Spreading	Mound-like	Globe	Columnar	Upright Spreading	Leggy	Closed to Base	Fine	Medium	Coarse	Green	Variegated	Grey-Green	Purple, Red	Silver	Yellow
Dogwood																				
Pagoda Dogwood				•					•	•				•	•					
Red-Osier Dogwood			•			•					•			•	•					
Flaviramea			•				•				•			•	•					
Isanti			•			•					•			•	•					
Kelseyi		•							•					•	•					
White Gold			•			•					•		•			•				
Rugose-Leaved Dogwood			•						•	•				•	•					
Tatarian Dogwood			•				•							•	•					
Argenteo-marginata		•				•					•		•			•				
Aurea		•				•								•						•
Gouchaultii		•				•					•		•				•			
Kesselringii		•					•				•			•				•		
Sibirica		•					•				•			•	•					
Dyers Greenweed		•							•			•			•					
Elder																				
American Elder		•							•	•				•	•					
Golden Elder			•			•					•			•						•
Red Elder			•			•	•							•	•					
Goldenlocks		•				•					•		•							•
Plumosa Aurea			•			•					•			•						•
Redman			•						•					•	•					
Sutherland Golden			•			•					•			•						•
False-Spirea																				
Ural False-Spirea			•						•			•			•					
Forsythia																				
Korean Goldenbells			•						•	•			•		•					
Northern Gold			•						•				•		•					
Ottawa		•							•				•		•					
Honeysuckle																				
Albert Thorn Honeysuckle		•			•							•			•					
Sakhalin Honeysuckle			•			•							•		•					
Sweetberry Honeysuckle			•			•				•			•		•					

Deciduous Shrubs

White	Violet, Purple	Red, Pink	Yellow	Fragrant	Ornamental Bark	Ornamental Fruit	Dry Conditions	Wet Conditions	Full Sun	Partial Shade	Sheltered Site	Salt Tolerant	Tender	Negative Characteristics	Limitations	Autumn Color	Winter Effect	Page Number	Plant Name
																			Dogwood
•										•	•							59	Pagoda Dogwood
•					•	•		•		•						•	•	60	Red-Osier Dogwood
•					•			•	•	•							•	60	Flaviramea
•					•					•						•	•	60	Isanti
								•	•				•					60	Kelseyi
•					•			•	•	•							•	60	White Gold
						•	•	•	•							•		60	Rugose-Leaved Dogwood
•					•	•			•							•	•	58	Tatarian Dogwood
							•	•	•									58	Argenteo-marginata
									•	•								59	Aurea
									•	•								59	Gouchaultii
					•				•	•								59	Kesselringii
•					•	•			•	•						•	•	59	Sibirica
			•				•		•									78	**Dyers Greenweed**
																			Elder
•				•		•												144	American Elder
•									•									144	Golden Elder
•						•			•	•								145	Red Elder
•									•									145	Goldenlocks
•									•									145	Plumosa Aurea
•						•				•	•							145	Redman
•									•									145	Sutherland Golden
																			False-Spirea
•									•	•								148	Ural False-Spirea
																			Forsythia
			•						•		•		•					74	Korean Goldenbells
			•						•		•		•					75	Northern Gold
			•						•		•		•					75	Ottawa
																			Honeysuckle
	•					•	•		•						•			91	Albert Thorn Honeysuckle
						•	•	•										91	Sakhalin Honeysuckle
•						•			•									90	Sweetberry Honeysuckle

Woody Ornamentals

Deciduous Shrubs

Plant Name	Size				Form							Texture			Foliage					
	Less than 30 cm	0.3–1 m	1–2 m	2–5 m	Low-Spreading	Mound-like	Globe	Columnar	Upright Spreading	Leggy	Closed to Base	Fine	Medium	Coarse	Green	Variegated	Grey-Green	Purple, Red	Silver	Yellow
Tartarian Honeysuckle				•					•	•			•		•					
Arnold Red				•					•	•			•		•					
Beaver Mor				•					•	•			•		•					
Carleton				•					•	•			•		•					
Morden Orange				•					•	•			•		•					
Vienna Honeysuckle		•					•				•		•		•					
Claveys Dwarf		•					•				•		•		•					
Miniglobe		•					•				•		•		•					
Zabels Honeysuckle				•					•		•		•		•					
Hydrangea																				
Pee Gee Hydrangea			•						•				•	•						
Praecox		•							•				•	•						
Snow Hills Hydrangea		•				•					•		•	•						
Annabelle		•				•					•		•	•						
Lilac																				
Chengtu Lilac				•					•				•	•						
Fountain				•					•				•	•						
Common Lilac				•					•		•		•	•						
Charm				•					•				•	•						
Congo				•					•	•			•	•						
Charles Joly				•					•	•			•	•						
Edith Cavell				•					•	•			•	•						
Gen. Pershing				•					•	•			•	•						
Katherine Havemeyer				•					•	•			•	•						
Ludwig Spaeth				•					•	•			•	•						
Madame Lemoine				•					•	•			•	•						
Montaigne				•					•	•			•	•						
Mrs. E. Wilmott				•					•	•			•	•						
Mrs. Edward Harding				•					•	•			•	•						
Hyacinth-Flowered Lilac				•					•	•			•	•						
Assessippi				•					•	•			•	•						
Minnehaha				•					•	•			•	•						
Pocahontas				•					•	•			•	•						
Sister Justina				•					•	•			•	•						
Swarthmore				•					•	•			•	•						

Deciduous Shrubs

Flowers				Special Features														Page Number	Plant Name
White	Violet, Purple	Red, Pink	Yellow	Fragrant	Ornamental Bark	Ornamental Fruit	Dry Conditions	Wet Conditions	Full Sun	Partial Shade	Sheltered Site	Salt Tolerant	Tender	Negative Characteristics	Limitations	Autumn Color	Winter Effect		
		•				•			•					•				91	Tartarian Honeysuckle
		•				•			•					•				91	Arnold Red
		•				•			•					•				91	Beaver Mor
		•				•			•					•				92	Carleton
		•				•			•					•				92	Morden Orange
									•									92	Vienna Honeysuckle
									•									92	Claveys Dwarf
									•									92	Miniglobe
	•								•		•							90	Zabels Honeysuckle
Hydrangea																			
		•							•	•	•							82	Pee Gee Hydrangea
		•							•	•	•							82	Praecox
•										•	•							82	Snow Hills Hydrangea
•										•	•							82	Annabelle
Lilac																			
		•							•									156	Chengtu Lilac
		•							•									156	Fountain
	•			•					•					•				157	Common Lilac
		•		•					•									158	Charm
	•			•					•									158	Congo
		•		•					•									158	Charles Joly
•				•					•									158	Edith Cavell
		•		•					•									158	Gen. Pershing
	•			•					•									158	Katherine Havemeyer
		•		•					•									158	Ludwig Spaeth
•				•					•									158	Madame Lemoine
		•		•					•									158	Montaigne
•				•					•									158	Mrs. E. Wilmott
		•		•					•									158	Mrs. Edward Harding
	•			•					•									153	Hyacinth-Flowered Lilac
	•			•					•									153	Assessippi
	•			•					•									153	Minnehaha
	•			•					•									153	Pocahontas
•				•					•									153	Sister Justina
		•		•					•									153	Swarthmore

Deciduous Shrubs

Plant Name	Less than 30 cm	0.3–1 m	1–2 m	2–5 m	Low-Spreading	Mound-like	Globe	Columnar	Upright Spreading	Leggy	Closed to Base	Fine	Medium	Coarse	Green	Variegated	Grey-Green	Purple, Red	Silver	Yellow
Josiflexa Lilac				•					•	•				•	•					
Bellicent				•					•	•				•	•					
Guinevere				•					•	•				•	•					
Lynette				•					•	•				•			•			
Korean Early Lilac				•					•	•				•	•					
Late Lilac				•			•				•			•	•					
Manchurian Lilac				•					•	•				•	•					
Meyer Lilac			•				•				•		•		•					
Persian Lilac			•				•				•		•		•					
Preston Lilac				•			•				•			•	•					
Coral				•			•				•			•	•					
Donald Wyman				•			•				•			•	•					
Isabella				•			•				•			•	•					
James McFarlane				•			•				•			•	•					
Jessica				•			•				•			•	•					
Minuet				•			•				•			•	•					
Miss Canada			•				•							•	•					
Rouen Lilac			•						•				•		•					
Mockorange Hybrids:																				
Audrey				•			•						•	•						
Marjorie				•			•						•	•						
Minnesota Snowflake				•					•	•			•	•						
Patricia		•					•						•	•						
Purity			•				•						•	•						
Sylvia			•				•						•	•						
Sweet Mockorange				•					•	•			•		•					
Waterton Mockorange			•						•		•		•		•					
Ninebark																				
Common Ninebark				•			•			•			•	•						
Dart's Gold		•					•				•		•							•
Luteus			•				•			•			•							•
Oleaster																				
Wolf-Willow, Silver Berry			•						•	•			•						•	
Oregon-Grape		•					•			•			•		•					

Deciduous Shrubs

White	Violet, Purple	Red, Pink	Yellow	Fragrant	Ornamental Bark	Ornamental Fruit	Dry Conditions	Wet Conditions	Full Sun	Partial Shade	Sheltered Site	Salt Tolerant	Tender	Negative Characteristics	Limitations	Autumn Color	Winter Effect	Page Number	Plant Name
	•			•					•									154	Josiflexa Lilac
		•		•					•									154	Bellicent
	•			•					•									154	Guinevere
		•		•					•									154	Lynette
	•			•					•									155	Korean Early Lilac
	•								•									157	Late Lilac
	•			•					•	•								157	Manchurian Lilac
		•		•					•	•								154	Meyer Lilac
	•			•					•									155	Persian Lilac
		•							•									155	Preston Lilac
		•							•									155	Coral
	•								•									155	Donald Wyman
		•							•									155	Isabella
		•							•									155	James McFarlane
	•								•									155	Jessica
		•		•					•									155	Minuet
		•		•					•									155	Miss Canada
		•		•					•									153	Rouen Lilac
																			Mockorange Hybrids:
•				•					•	•								102	Audrey
•				•					•	•								102	Marjorie
•				•					•	•								102	Minnesota Snowflake
•				•					•	•								102	Patricia
•				•					•	•								102	Purity
•				•					•									102	Sylvia
•				•					•	•								102	Sweet Mockorange
•									•	•								102	Waterton Mockorange
																			Ninebark
•						•			•	•								104	Common Ninebark
•						•			•									104	Dart's Gold
•						•			•									104	Luteus
																			Oleaster
			•			•			•									70	Wolf-Willow, Silver Berry
			•			•			•	•				•	•			94	**Oregon-Grape**

Woody Ornamentals

Deciduous Shrubs

Plant Name	Size				Form							Texture			Foliage					
	Less than 30 cm	0.3 – 1 m	1–2 m	2–5 m	Low-Spreading	Mound-like	Globe	Columnar	Upright Spreading	Leggy	Closed to Base	Fine	Medium	Coarse	Green	Variegated	Grey-Green	Purple, Red	Silver	Yellow
Plum, Cherry, Chokecherry																				
Chinese Bush Cherry		•					•				•		•		•					
Chokecherry				•			•		•	•				•	•					
Boughen's Chokeless				•			•		•	•				•	•					
Boughen's Yellow				•			•			•				•	•					
Flowering Plum				•			•				•		•		•					
Double Flowering Plum				•			•				•		•		•					
Mongolian Cherry			•				•				•		•		•					
Nanking Cherry			•						•		•		•		•					
Prairie Almond			•				•				•		•		•					
Purple-Leaved Sandcherry × Plum Hybrid		•							•		•		•					•		
Russian Almond		•					•				•				•					
Sandcherry		•			•								•				•			
Western Sandcherry			•						•		•		•		•					
Pricklyspine				•			•				•			•	•					
Prinsepia																				
Cherry Prinsepia			•				•						•		•					
Rest Harrow, Ononis		•					•					•			•					
Rose, Shrub Rose																				
Altai Rose, Altai Scot's Rose			•						•	•		•			•					
Yellow Altai			•						•	•		•			•					
var. *altaica*			•							•		•			•					
Austrian Brier Rose			•						•			•			•					
Bicolor			•						•			•			•					
Harisonii			•						•		•	•			•					
Persiana			•						•		•	•			•					
French Rose			•						•		•		•		•					
Grandiflora				•					•		•		•		•					
Prickly Rose		•							•		•		•		•					
Red-Leaf Rose			•						•		•		•						•	
Rugosa Rose			•				•				•			•	•					
Blanc Double de Coubert			•				•				•			•	•					
F.J. Grootendorst		•							•		•	•			•					
Hansa			•						•				•		•					
Martin Frobisher			•						•				•		•					

Deciduous Shrubs

White	Violet, Purple	Red, Pink	Yellow	Fragrant	Ornamental Bark	Ornamental Fruit	Dry Conditions	Wet Conditions	Full Sun	Partial Shade	Sheltered Site	Salt Tolerant	Tender	Negative Characteristics	Limitations	Autumn Color	Winter Effect	Page Number	Plant Name
•						•			•	•									**Plum, Cherry, Chokecherry**
•							•									•		121	Chinese Bush Cherry
•					•	•			•							•		126	Chokecherry
•						•			•							•		126	Boughen's Chokeless
•						•			•							•		126	Boughen's Yellow
•									•	•								126	Flowering Plum
	•								•	•			•					126	Double Flowering Plum
•						•			•					•				121	Mongolian Cherry
•						•			•	•								125	Nanking Cherry
		•							•									124	Prairie Almond
•						•			•				•					121	Purple-Leaved Sandcherry × Plum Hybrid
		•							•						•			125	Russian Almond
•								•	•									125	Sandcherry
•									•					•	•			120	Western Sandcherry
	•								•	•								35	**Pricklyspine**
																			Prinsepia
			•		•				•							•		120	Cherry Prinsepia
	•						•		•									98	**Rest Harrow, Ononis**
																			Rose, Shrub Rose
•									•						•			139	Altai Rose, Altai Scot's Rose
			•						•						•			139	Yellow Altai
•									•						•			139	var. *altaica*
			•						•						•			137	Austrian Brier Rose
	•	•							•									137	Bicolor
			•						•									137	Harisonii
			•						•					•				137	Persiana
	•								•									138	French Rose
	•								•									138	Grandiflora
		•							•					•				135	Prickly Rose
	•								•									138	Red-Leaf Rose
	•								•									138	Rugosa Rose
•									•									138	Blanc Double de Coubert
	•								•									138	F.J. Grootendorst
	•								•									138	Hansa
	•								•									138	Martin Frobisher

Deciduous Shrubs

Plant Name	Size				Form							Texture			Foliage					
	Less than 30 cm	0.3 – 1 m	1–2 m	2–5 m	Low-Spreading	Mound-like	Globe	Columnar	Upright Spreading	Leggy	Closed to Base	Fine	Medium	Coarse	Green	Variegated	Grey-Green	Purple, Red	Silver	Yellow
Smooth Rose		•							•	•		•			•					
Betty Bland			•						•		•		•		•					
Therese Bugnet				•			•			•			•		•					
Sunshine Rose	•				•								•		•					
Adelaide Hoodless		•								•			•		•					
Assiniboine		•								•			•		•					
Cuthbert Grant		•								•			•		•					
Morden Amorette		•								•			•		•					
Morden Cardinette		•							•				•		•					
Morden Centennial			•						•				•		•					
Morden Ruby			•							•			•		•					
Salt-tree				•					•	•			•				•			
Sea-buckthorn				•					•	•		•							•	
Saskatoon																				
Saskatoon				•					•	•			•		•					
Altaglow				•					•	•			•		•					
Honeywood				•					•	•			•		•					
Northline				•					•		•		•		•					
Pembina				•					•	•			•		•					
Smoky				•					•		•		•		•					
Thiessen				•				•					•		•					
Spirea																				
Billiard Spirea		•							•				•		•					
Bridalwreath Spirea			•				•			•			•		•					
Dwarf Pink Spirea		•				•					•		•		•					
Anthony Waterer		•				•					•		•		•					
Crispa		•							•		•		•		•					
Dart's Red		•					•				•		•						•	
Froebeli		•				•					•		•		•					
Gold Flame		•				•					•		•							•
Garland Spirea			•				•				•	•					•			
Japanese Spirea		•					•						•		•					
Alpina	•				•							•			•					
Goldmound		•				•						•								•
Little Princess		•			•						•	•			•					

Deciduous Shrubs

Flowers				Special Features													Page Number	Plant Name
White	Violet, Purple	Red, Pink	Yellow	Fragrant	Ornamental Bark	Ornamental Fruit	Dry Conditions	Wet Conditions	Full Sun	Partial Shade	Sheltered Site	Salt Tolerant	Tender	Negative Characteristics	Limitations	Autumn Color	Winter Effect		
		•							•	•								136	Smooth Rose
		•							•									137	Betty Bland
		•							•									137	Therese Bugnet
		•				•			•									135	Sunshine Rose
		•		•					•		•							136	Adelaide Hoodless
	•								•		•							136	Assiniboine
		•							•		•		•					136	Cuthbert Grant
		•							•		•		•					136	Morden Amorette
		•							•		•							136	Morden Cardinette
		•							•									136	Morden Centennial
		•							•		•		•					136	Morden Ruby
	•								•			•						79	**Salt-tree**
						•	•		•			•						81	**Sea-buckthorn**
																			Saskatoon
•						•			•		•					•		47	Saskatoon
•						•			•		•					•		47	Altaglow
•						•			•		•							47	Honeywood
•						•			•		•							47	Northline
•						•			•		•							47	Pembina
•						•			•		•							47	Smoky
•						•			•		•	•						47	Thiessen
																			Spirea
		•							•	•								150	Billiard Spirea
•									•									152	Bridalwreath Spirea
		•							•		•							151	Dwarf Pink Spirea
		•							•									151	Anthony Waterer
		•							•		•							151	Crispa
		•							•		•							151	Dart's Red
		•							•		•							151	Froebeli
		•							•		•						•	151	Gold Flame
•									•		•							150	Garland Spirea
		•							•		•							151	Japanese Spirea
		•							•		•					•		151	Alpina
		•							•		•							151	Goldmound
		•							•		•						•	151	Little Princess

Deciduous Shrubs

Plant Name	Size				Form							Texture			Foliage					
	Less than 30 cm	0.3–1 m	1–2 m	2–5 m	Low-Spreading	Mound-like	Globe	Columnar	Upright Spreading	Leggy	Closed to Base	Fine	Medium	Coarse	Green	Variegated	Grey-Green	Purple, Red	Silver	Yellow
Oriental Spirea			•				•				•	•			•					
Three-Lobed Spirea		•					•				•		•		•					
Sumac																				
Lemonade Sumac			•		•						•		•		•					
Smooth Sumac				•					•	•				•	•					
Midi			•				•				•		•	•	•					
Mini		•					•				•		•	•	•					
Staghorn Sumac				•					•	•				•	•					
Tamarisk																				
Amur Tamarisk				•					•	•		•			•					
Viburnum																				
High Bush-Cranberry				•					•	•			•		•					
Andrews			•				•				•		•		•					
Compactum		•					•				•		•		•					
Garry Pink				•			•				•		•		•					
Arrowwood				•					•		•		•		•					
Guelder Rose			•						•	•			•		•					
Compactum		•					•				•		•		•					
Nanum	•						•				•	•			•					
Sterile		•							•		•		•		•					
Xanthocarpum				•					•	•			•		•					
Nannyberry				•					•	•			•		•					
Sargent's Viburnum				•					•	•			•		•					
Wayfaring Tree				•			•				•			•			•			
Willow																				
Blue Fox Willow		•					•				•	•					•			
Coyote Willow			•						•			•								•
Purple Osier Willow		•				•					•	•					•			
Sandbar Willow				•					•	•		•					•			
Woolly Willow		•			•								•				•			
Woadwaxen																				
Dyer's Greenweed		•							•			•	•		•					
Rossica		•				•						•	•		•					
Lydia Woadwaxen		•				•							•		•					

Deciduous Shrubs

White	Violet, Purple	Red, Pink	Yellow	Fragrant	Ornamental Bark	Ornamental Fruit	Dry Conditions	Wet Conditions	Full Sun	Partial Shade	Sheltered Site	Salt Tolerant	Tender	Negative Characteristics	Limitations	Autumn Color	Winter Effect	Page Number	Plant Name
•									•									152	Oriental Spirea
•									•									152	Three-Lobed Spirea
																			Sumac
						•	•		•									133	Lemonade Sumac
						•	•		•					•	•			132	Smooth Sumac
						•			•							•		133	Midi
						•			•							•		133	Mini
						•	•		•							•		133	Staghorn Sumac
																			Tamarisk
		•					•		•									158	Amur Tamarisk
																			Viburnum, Bush Cranberry
•						•		•	•	•						•	•	167	High Bush-Cranberry
•						•			•	•						•	•	167	Andrews
									•							•		167	Compactum
		•				•		•	•							•		167	Garry Pink
•						•			•	•						•		166	Arrowwood
•									•	•	•							165	Guelder Rose
•						•				•	•							166	Compactum
									•	•	•							166	Nanum
•									•	•		•						166	Sterile
•						•			•	•						•		166	Xanthocarpum
•									•	•						•		165	Nannyberry
•						•			•	•						•	•	166	Sargent's Viburnum
•						•			•	•						•		165	Wayfaring Tree
																			Willow
									•	•								141	Blue Fox Willow
							•	•	•	•								142	Coyote Willow
					•			•	•									143	Purple Osier Willow
								•	•									142	Sandbar Willow
									•	•			•					142	Woolly Willow
																			Woadwaxen
			•						•									78	Dyer's Greenweed
			•						•									78	Rossica
			•						•									78	Lydia Woadwaxen

Woody Ornamentals

Deciduous Trees

Plant Name	Under 5 m	5–10 m	10–15 m	15–20 m	20–30 m	High-headed	Low-headed	Globular	Weeping	Upright Oval	Upright Spreading	Columnar	Pyramidal	Excurrent	Decurrent	Open	Dense	Fine	Coarse
Alder																			
Mountain Alder		•					•			•					•		•		•
Red Alder		•					•				•				•	•			•
Amur Maackia	•						•			•					•	•		•	
Ash																			
Black Ash			•			•								•	•		•		•
Fallgold			•			•								•	•		•		•
Green Ash				•		•				•					•		•		•
Patmore				•		•				•					•		•		•
Manchurian Ash			•				•			•					•		•		•
White Ash			•				•			•					•		•		•
Autumn Blaze			•				•			•					•		•		•
Basswood, Linden																			
Basswood				•		•								•	•		•		•
Dropmore Linden				•		•								•	•		•		•
Wascana				•		•								•	•		•		•
Little-Leaf Linden			•					•						•	•		•		•
Greenspire			•			•								•	•		•		•
Morden			•			•								•	•		•		•
Mongolian Linden		•					•							•	•		•		•
Birch																			
Chinese Paper Birch		•					•			•					•	•		•	
Dahurian Birch			•				•			•					•	•		•	
European Birch			•				•			•					•	•		•	
Fastigiata			•				•					•		•		•		•	
Gracilis				•			•		•						•	•		•	
Youngii	•						•		•						•	•		•	
Paper Birch		•					•			•					•	•		•	
Chickadee		•					•			•					•	•		•	
River Birch		•					•			•					•	•			•
Water Birch			•			•	•								•	•			•
Buckeye, Horse Chestnut																			
Horse Chestnut			•			•	•								•		•		•
Ohio Buckeye		•				•				•					•		•		•

Deciduous Trees

Foliage				Flowers						Special Features											Page Number	Plant Name
Green	Purple, Red	Silver	Yellow	White	Red	Pink	Yellow	Fragrant		Ornamental Bark	Ornamental Fruit	Dry Conditions	Wet Conditions	Full Sun	Sheltered	Salt Tolerant	Hardiness (?)	Negative Characteristics	Limitations	Autumn Color	Winter Effect		
																							Alder
•													•	•								42	Mountain Alder
•													•	•								42	Red Alder
•			•											•								93	**Amur Maackia**
																							Ash
•													•	•								76	Black Ash
•													•	•						•		76	Fallgold
•														•						•		77	Green Ash
•														•						•		77	Patmore
•														•								76	Manchurian Ash
•														•	•					•		75	White Ash
•														•	•					•		75	Autumn Blaze
																							Basswood, Linden
•							•	•						•						•		161	Basswood
•							•	•						•						•		162	Dropmore Linden
•							•	•						•						•		162	Wascana
•							•	•						•						•		161	Little-Leaf Linden
•							•	•						•						•		162	Greenspire
•							•	•						•						•		162	Morden
•							•	•						•						•		162	Mongolian Linden
																							Birch
•										•				•						•	•	49	Chinese Paper Birch
•										•				•						•	•	49	Dahurian Birch
•										•				•						•	•	51	European Birch
•										•				•	•					•	•	51	Fastigiata
•										•				•						•	•	51	Gracilis
•										•				•			•			•	•	51	Youngii
•										•				•						•	•	50	Paper Birch
•										•				•						•	•	51	Chickadee
•										•			•									50	River Birch
•										•										•	•	50	Water Birch
																							Buckeye, Horse Chestnut
•				•									•	•				•				41	Horse Chestnut
•				•									•	•						•		41	Ohio Buckeye

Woody Ornamentals

Deciduous Trees

Plant Name	Under 5 m	5–10 m	10–15 m	15–20 m	20–30 m	High-headed	Low-headed	Globular	Weeping	Upright Oval	Upright Spreading	Columnar	Pyramidal	Excurrent	Decurrent	Open	Dense	Fine	Coarse
Butternut, Walnut																			
Black Walnut			•				•			•					•	•			•
Butternut			•				•				•				•	•			•
Corktree																			
Amur Corktree		•					•				•				•	•			•
Crab Apple																			
Red Jade Crab Apple	•						•	•							•		•		•
Rosybloom Crab Apples																			
Almey		•					•								•		•		•
Arctic Dawn	•						•	•							•		•		•
Hopa		•					•								•		•		•
Kelsey	•						•								•		•		•
Pygmy	•						•	•						•			•		•
Red Splendor	•						•		•						•		•		•
Royalty		•					•				•				•		•		•
Rudolph	•						•	•							•		•		•
Selkirk	•						•	•							•		•		•
Strathmore		•					•				•				•	•			•
Thunderchild	•						•	•							•	•			•
Siberian Crab Apple		•					•	•							•		•		•
Columnaris	•						•					•		•			•		•
Snowcap	•						•			•					•		•		•
Tanner Crab Apple	•						•			•					•	•			•
Elm																			
American Elm						•	•			•					•		•		•
Beaverlodge				•		•				•					•		•		•
Brandon				•		•				•					•		•		•
Japanese Elm			•			•				•					•		•		•
Manchurian Elm				•		•				•					•		•	•	
Hackberry																			
Delta Hackberry				•		•				•					•		•		•
Hawthorn																			
Arnold Hawthorn	•						•	•							•		•		•
Fleshy Hawthorn	•						•	•							•		•		•

Deciduous Trees

Foliage				Flowers					Special Features												Page Number	Plant Name
Green	Purple, Red	Silver	Yellow	White	Red	Pink	Yellow	Fragrant	Ornamental Bark	Ornamental Fruit	Dry Conditions	Wet Conditions	Full Sun	Sheltered	Salt Tolerant	Hardiness (?)	Negative Characteristics	Limitations	Autumn Color	Winter Effect		
																						Butternut, Walnut
•													•						•	•	83	Black Walnut
•													•						•	•	83	Butternut
																						Corktree
•										•	•								•		101	Amur Corktree
																						Crab Apple
•						•				•			•						•	•	97	Red Jade Crab Apple
																					95	Rosybloom Crab Apples
•						•				•			•	•						•	95	Almey
•						•				•			•	•							95	Arctic Dawn
•						•							•	•							95	Hopa
	•				•								•								95	Kelsey
	•				•								•	•							96	Pygmy
•						•				•			•	•						•	96	Red Splendor
	•				•								•	•			•	•			96	Royalty
•					•								•	•							96	Rudolph
•						•				•			•	•						•	96	Selkirk
	•				•								•	•							96	Strathmore
	•				•								•	•							96	Thunderchild
•			•							•			•	•						•	96	Siberian Crab Apple
•			•										•	•							96	Columnaris
•			•										•	•						•	96	Snowcap
•			•										•	•							95	Tanner Crab Apple
																						Elm
•													•				•	•			163	American Elm
•																•					163	Beaverlodge
•																•	•				163	Brandon
•												•				•	•				164	Japanese Elm
•												•				•	•				164	Manchurian Elm
																						Hackberry
•													•								55	Delta Hackberry
																						Hawthorn
•				•						•			•			•					65	Arnold Hawthorn
•				•						•			•			•					66	Fleshy Hawthorn

Deciduous Trees

Plant Name	Size					Form								Habit		Canopy		Texture	
	Under 5 m	5–10 m	10–15 m	15–20 m	20–30 m	High-headed	Low-headed	Globular	Weeping	Upright Oval	Upright Spreading	Columnar	Pyramidal	Excurrent	Decurrent	Open	Dense	Fine	Coarse
Morden Hybrid Hawthorn																			
Snowbird	•									•					•		•		•
Toba	•										•				•		•		•
Lilac																			
Japanese Tree Lilac		•				•					•				•				•
Maidenhair Tree		•				•					•				•	•		•	
Maple																			
Amur Maple	•					•	•								•		•		
var. *seminowii*	•					•	•								•		•		
Manitoba Maple, Box-Elder		•				•				•					•		•		•
Baron		•				•				•					•		•		•
Mountain Maple	•					•					•				•	•			•
Norway Maple		•				•				•					•		•		•
Crimson King		•				•				•					•		•		•
Schwedleri		•				•				•					•		•		•
Rocky Mountain Maple	•					•					•				•	•			•
Silver Maple				•		•				•					•		•		•
Northline				•		•				•					•		•		•
Sugar Maple			•			•				•					•		•		•
Tartarian Maple		•				•				•					•		•		
Mountain-Ash																			
American Mountain-Ash		•				•				•					•		•	•	
European Mountain-Ash		•				•	•								•		•	•	
Fastigiata		•				•						•		•			•	•	
Western Mountain-Ash		•				•	•								•		•	•	
Oak																			
Bur Oak			•			•				•					•		•		•
Mongolian Oak			•			•				•					•		•		
Northern Red Oak			•			•						•			•		•		•
Scarlet Oak		•				•				•					•	•			
White Oak				•		•				•					•		•		
Oleaster																			
Russian-Olive		•						•			•				•		•	•	
Pear																			
Ussurian Pear	•					•				•					•		•		•

Deciduous Trees

Foliage				Flowers					Special Features											Page Number	Plant Name
Green	Purple, Red	Silver	Yellow	White	Red	Pink	Yellow	Fragrant	Ornamental Bark	Ornamental Fruit	Dry Conditions	Wet Conditions	Full Sun	Sheltered	Salt Tolerant	Hardiness (?)	Negative Characteristics	Limitations	Autumn Color	Winter Effect		
																					65	Morden Hybrid Hawthorn
•				•						•			•								66	Snowbird
•						•				•						•		•			66	Toba
																						Lilac
•				•									•	•							156	Japanese Tree Lilac
•													•		•	•			•		80	**Maidenhair Tree**
																						Maple
•				•						•			•						•		36	Amur Maple
•				•						•			•						•		36	var. *seminowii*
•				•									•				•				37	Manitoba Maple, Box-Elder
•				•									•				•				37	Baron
•				•						•			•	•					•		39	Mountain Maple
•													•	•							37	Norway Maple
	•												•	•							38	Crimson King
	•												•	•							38	Schwedleri
•													•						•		36	Rocky Mountain Maple
•					•								•						•		38	Silver Maple
•					•								•						•		38	Northline
•													•	•					•		39	Sugar Maple
•								•	•				•						•		39	Tartarian Maple
																						Mountain-Ash
•				•			•			•			•	•					•	•	148	American Mountain-Ash
•				•			•			•			•	•					•	•	149	European Mountain-Ash
•				•			•			•			•	•					•	•	149	Fastigiata
•				•			•			•			•	•					•	•	149	Western Mountain-Ash
																						Oak
•													•						•		130	Bur Oak
•														•							131	Mongolian Oak
•											•		•						•		129	Northern Red Oak
•											•		•						•		130	Scarlet Oak
•							•														129	White Oak
																						Oleaster
		•				•	•				•		•	•							70	Russian-Olive
																						Pear
•				•									•	•			•				128	Ussurian Pear

Deciduous Trees

Plant Name	Size					Form								Habit		Canopy		Texture	
	Under 5 m	5–10 m	10–15 m	15–20 m	20–30 m	High-headed	Low-headed	Globular	Weeping	Upright Oval	Upright Spreading	Columnar	Pyramidal	Excurrent	Decurrent	Open	Dense	Fine	Coarse
Poplar																			
Balsam Poplar					•	•				•					•	•			•
Berlin Poplar			•			•				•				•			•		•
European Columnar Aspen		•										•		•			•		•
Northwest Poplar					•	•								•			•		•
Plains' Cottonwood					•	•					•				•		•		•
Russian Poplar			•			•				•				•			•		•
Dunlop				•		•							•				•		•
Griffin			•							•				•			•		•
Walker			•			•									•		•		•
Silver Poplar		•				•				•				•			•		•
Tower Poplar		•										•		•			•		•
Trembling Aspen				•		•				•				•	•			•	
Plum, Cherry, Chokecherry																			
Amur Cherry			•			•				•				•			•		•
Canada Plum	•					•	•								•		•		•
Chokecherry	•					•		•							•		•		•
Copper Shubert	•					•					•				•	•			•
Mini-Shubert	•					•				•				•			•		•
Shubert	•					•						•			•		•		•
Sharon	•					•									•		•		•
Mayday Tree, European Bird Cherry			•					•							•	•			•
Muckle Plum	•					•				•					•	•			•
Pincherry		•				•				•					•	•			•
Jumping Pound	•								•						•		•		•
Mary Liss		•				•				•					•		•		•
Stockton		•				•				•					•		•		•
Willow																			
Daphne Willow	•					•				•					•	•	•		
Fangstad Willow	•								•						•		•		
Laurel-Leaf Willow			•			•				•					•		•		•
Pussy Willow	•					•	•								•				
Sharp-Leaf Willow		•				•	•								•		•	•	
White Willow				•		•	•			•					•		•		•
Golden Willow		•				•	•			•					•		•	•	
Siberian White Willow			•							•					•		•		•

Deciduous Trees

Foliage				Flowers					Special Features												Page Number	Plant Name
Green	Purple, Red	Silver	Yellow	White	Red	Pink	Yellow	Fragrant	Ornamental Bark	Ornamental Fruit	Dry Conditions	Wet Conditions	Full Sun	Sheltered	Salt Tolerant	Hardiness (?)	Negative Characteristics	Limitations	Autumn Color	Winter Effect		
																						Poplar
•												•				•			•		114	Balsam Poplar
•													•			•					114	Berlin Poplar
•													•			•					117	European Columnar Aspen
•													•			•			•		115	Northwest Poplar
•												•	•			•			•		115	Plains' Cottonwood
•													•			•			•		116	Russian Poplar
•													•			•			•		116	Dunlop
•													•			•			•		116	Griffin
•													•			•			•		116	Walker
•													•			•	•		•		114	Silver Poplar
•													•			•					115	Tower Poplar
•													•			•			•		117	Trembling Aspen
																						Plum, Cherry, Chokecherry
•				•					•				•				•	•	•	•	122	Amur Cherry
•				•									•	•							122	Canada Plum
•				•						•			•						•		126	Chokecherry
	•												•								127	Copper Shubert
	•												•								127	Mini-Shubert
	•			•									•								127	Sharon
•										•			•								127	Shubert
•				•				•					•								123	Mayday Tree, European Bird Cherry
•						•							•	•							123	Muckle Plum
•				•							•	•	•						•		124	Pin Cherry
•				•								•	•						•		124	Jumping Pound
•				•							•		•						•		124	Mary Liss
•				•						•			•						•		124	Stockton
																						Willow
•				•					•			•	•	•							141	Daphne Willow
•												•	•				•	•			142	Fangstad Willow
•													•								143	Laurel-Leaf Willow
•		•											•								142	Pussy Willow
•													•								140	Sharp-Leaf Willow
													•								140	White Willow
•									•				•	•							141	Golden Willow
		•							•				•								141	Siberian White Willow

Woody Ornamentals

Groundcover Plants

Plant Name	Size				Form				Texture			Foliage					
	10 cm	20 cm	30 cm	40 cm	Hummock	Low Spreading	Upright Spreading	Ground-hugging	Fine	Medium	Coarse	Green	Variegated	Purple	Red	Silver	Yellow
Arctic Phlox	•							•	•			•					
Baltic Ivy																	
Birdsfoot Trefoil	•					•				•		•					
Bishop's Goutweed			•			•				•			•				
Blue Sheep's Fescue		•				•			•							•	
Broom																	
Purple Broom			•			•				•		•					
Rock Garden Broom	•						•	•				•					
Bunchberry	•					•				•		•					
Cinquefoil, Three-Toothed		•				•			•			•					
Cliff Green, Mountain Lover																	
Cliff Green		•				•				•		•					
Oregon-Boxwood			•		•					•		•					
Creeping Oregon-Grape	•					•					•	•					
Crown Vetch			•		•				•			•					
Kinnikinik, Bearberry																	
Alpine Bearberry	•						•	•				•					
Kinnikinik	•					•				•		•					
Vancouver Jade	•					•				•		•					
Pachysandra																	
Japanese Spurge			•				•			•		•					
Periwinkle, Barvenok																	
Periwinkle	•					•				•		•					
Deciduous Periwinkle	•							•	•			•					
Polygonum				•		•				•		•					
Pussytoes	•							•	•							•	
Running Strawberrybush		•				•				•		•					
Sweet Woodruff			•				•			•		•					
Large-Leaf Saxifrage, Siberian-Hyacinth			•			•					•	•					
Windflower Anemone			•				•			•		•					
Yellow-Flowered Strawberry	•							•		•		•					

Groundcover Plants

White	Purple, Red	Pink	Yellow	Blue	Ornamental Fruit	Dry Conditions	Wet Conditions	Full Sun	Partial Shade	Full Shade	Sheltered	Unsheltered	Salt Tolerant	Tender	Hardiness (?)	Negative Characteristics	Limitations	Autumn Color	Page Number	Plant Name
		•						•											103	Arctic Phlox
									•		•				•				80	Baltic Ivy
		•						•			•								93	Birds-foot Trefoil
•									•							•	•		40	Bishop's Goutweed
						•	•		•								•		74	Blue Sheep's Fescue
																				Broom
	•							•			•								67	Purple Broom
		•						•			•								67	Rock-Garden Broom
•					•			•												Bunchberry
•								•	•									•	119	Cinquefoil, Three-Toothed
																				Cliff-Green, Mountain Lover
									•	•		•							100	Cliff-Green
									•	•							•		100	Oregon-Boxwood
		•	•					•	•								•		94	Creeping Oregon-Grape
	•							•											61	Crown Vetch
•					•			•	•			•					•		113	Fleeceflower
																				Kinnikinik, Bearberry
•					•			•	•		•						•		45	Alpine Bearberry
		•			•			•	•		•								46	Kinnikinik
		•			•			•	•		•								46	Vancouver Jade
																				Spurge
•						•				•	•								98	Japanese Spurge
																				Periwinkle, Barvenok
				•					•	•	•								168	Bowles Periwinkle
				•					•	•									168	Deciduous Periwinkle
•						•		•											44	Pussytoes
					•			•	•										73	Running Strawberrybush
•										•									77	Sweet Woodruff
		•						•	•		•								48	Saxifrage, Large-Leaf, Siberian-Hyacinth
•								•	•			•							44	Windflower Anemone
			•		•			•			•								69	Yellow-Flowered Strawberry

Vines and Climbers

	Size		Texture		Flowers							Special Features							Page Number
	2–5 m	5–10 m	Medium	Coarse	White	Red	Pink	Yellow	Blue	Violet	Purple	Ornamental Fruit	Well-drained Soil	Cool Root Environment	Full Sun	Partial Shade	Autumn Color	Support Needed	
Bittersweet																			
American Bittersweet	•			•								•	•		•		•	•	54
Clematis, Virgin's Bower																			
Bigpetal Clematis	•			•					•				•	•				•	56
Golden Virgin's-Bower	•			•				•				•		•	•			•	57
Jackman Clematis	•			•							•			•	•			•	56
Western Blue Clematis	•			•					•			•		•	•	•		•	58
Clematis Macropetala **Hybrids**																			
Rosy O'Grady	•			•			•							•	•			•	57
White Swan	•			•	•									•	•			•	57
Other *Clematis* Hybrids																			
Blue Boy	•			•					•					•	•			•	57
Golden Cross			•		•			•						•	•			•	57
Grace	•	•		•										•	•	•		•	57
Nellie Moser	•			•			•						•	•	•		•	•	57
Pamela	•	•		•										•	•	•		•	57
Western White Clematis	•	•		•	•							•		•	•			•	57
Grape																			
Riverbank Grape		•		•								•			•		•	•	169
Beta		•		•								•			•		•	•	169
Valiant		•		•								•			•		•	•	169
Honeysuckle																			
Dropmore Scarlet Trumpet	•			•		•									•	•		•	90
Virginia Creeper																			
Engelman's Virginia Creeper		•		•											•		•		99
Virginia Creeper		•		•											•		•	•	99

Glossary

Achene A dry, one-compartment, one-seeded fruit that does not open when ripe.

Acuminate Abruptly sharp-pointed.

Adpressed Closely and flatly pressed together.

Acorn A nut with its basal portion partly covered by a cup-like structure.

Anther The pollen-bearing part of a stamen.

Apetalous Without petals.

Aril A tissue found covering the seeds of some plants, not as a seed coat but as an outgrowth from the point of attachment of the seed.

Awl-shape Sharp-pointed like a carpenter's awl.

Axil The upper angle formed by a leaf or a branch and the stem.

Berry A fleshy fruit developed from a single ovary and usually containing several seeds embedded in a fleshy pulp.

Bract A modified leaf from the axil of which a flower or flower cluster arises.

Calyx The outer whorl of flower parts. They are normally green.

Capsule A dry fruit of more than one chamber and opening at maturity.

Catkin A scaly spike of flowers of one sex.

Circumpolar The term used to describe a group of plants whose distribution circumscribes one of the polar regions.

Compound A term used to describe the leaf of a plant where the leaf consists of two or more separate parts (leaflets).

Compound Umbel A two-tiered umbel with the flowers borne on those in the upper tier.

Cordate Heart shaped.

Corolla The second whorl of flower parts. Individually these are known as petals. Normally they are the highly-colored parts of the flower.

Corymb A cluster of flowers in which the flower stalks arise from different parts of the stem. The top of the flower cluster is flat or nearly so and the first flowers to open are on the edges of the cluster.

Cultivar A cultivated variety as distinguished from a botanical variety.

Cyme A flat or slightly convex-topped flower cluster in which the central flower opens first.

Deciduous Dropping as with leaves; not persistent as with evergreen.

Decurrent The crown structure of a tree when the lateral branches tend to grow independently of the central or main stem.

Decumbent Prostrate at the base but erect or ascending elsewhere.

Deltoid Broadly triangular.

Dioecious The term used to describe a plant with imperfect flowers when the flowers of each sex are found on separate plants.

Dormancy/Quiesence A period of inactivity in plants that begins at the end of the period Dormancy/Rest and terminates when growing conditions resume in the spring. This period of inactivity is totally under control of the environment.

Dormancy/Rest A period of profound inactivity exhibited by plant growing points which begins just before the end of the growing season and ends sometime during early winter. This period of inactivity is controlled by the plant and is sometimes referred to as the annual rest-period.

Drupe A one seeded fleshy fruit in which the seed is contained within a hard bony shell.

Elipsoid Like an elipse.

Entire A term used to describe the margin of a leaf where the margin is without teeth or clefts.

Excurrent The term used to describe the crown structure of a tree when the central or main stem tends to exert control over growth of the lateral branches.

Exfoliating Peeling off in thin layers, as with the bark of most birch trees.

Fasciated Bound together in a bundle.

Fastigiate With branches erect and nearly parallel.

Gall An abnormal growth caused by an insect or disease.

Glaucous The term used to describe the color when a white or grey covering is found on stems or leaves.

Halophyte A plant that grows in salty soil.

Herbaceous Not woody; usually dying back to the ground each fall.

Imperfect A flower containing only male or female organs, not both.

Inflorescence An arrangement of flowers in a cluster.

Woody Ornamentals 207

Involucre A whorl of bracts surrounding a flower cluster.

Juvenile plant A plant exhibiting juvenile characteristics such as the inability to produce flowers throughout life.

Lanceolate Lance-shaped, at least 4 times as long as wide and broadest below the centre.

Leaflet The term applied to a single unit of a compound leaf.

Lenticel Small corky dots or horizontal lines on the bark of some shrubs or trees.

Monoecious Plants with unisexual flowers where the flowers of both sexes appear on the same plant.

Nut A single seeded fruit with a hard dry, outer coat.

Obovate The shape of an oval leaf where the widest part of the leaf is found beyond the mid-point.

Ovate The shape of a leaf where the base is broader than the tip; egg-shaped.

Panicle A branched flower cluster that is longer than wide.

Palmate The term used to describe the form of a compound leaf with five parts arranged like the extended fingers of the human hand. The term is also used to describe simple lobed leaves having five lobes.

Pedicel The stalk of a single flower.

Peduncle The stalk of a flower cluster.

Pendulous Hanging down.

Pericarp The fructified ovary.

Perfect A flower complete with both stamens and pistil.

Petaloid Petal-like.

Petiole A leaf stalk.

Pinnate A compound leaf where the leaflets are arranged on either side of a central axis.

Pistillate A flower with pistil or pistils but lacking stamens (a female flower)

Pistils The female portions of a flower.

Pith The central core of spongy cells in a stem.

Pod A dry dehiscent fruit; common to legumes.

Polygamous Bearing both perfect and unisexual flowers on the same plant.

Polygamodioecious Said of flowers; sometimes perfect, sometimes unisexual, the two forms borne on different plants.

Pome A fleshy fruit like the apple and pear in which the fleshy structure is derived from the base of the flower.

Prickle An outgrowth of the bark

Procumbent Growing along the ground.

Pubescent Clothed with soft short hairs.

Raceme An elongated flower cluster where the flowers are borne on short stalks from a single axis, with the oldest flowers occurring at the base.

Receptacle The base of a flower stalk bearing the flower organs or the base of a flower cluster bearing the florets.

Reflexed Bent sharply backwards.

Rosette A dense cluster of leaves on a very short stem or axis.

Rugose Wrinkled.

Semi-double A flower in which some, but not all, of the sexual organs have become petals or petal-like.

Sepal One of the separate parts of the calyx of a flower.

Sessile without a stalk.

Shrub A woody plant that branches from the base that is generally under four meters tall.

Simple A term used to describe a leaf consisting of one piece or the pistil of a flower when it is made up of a single carpel.

Sinus The space between the lobes of a lobed leaf.

Spike An elongated flower cluster where stalkless florets are borne along an axis.

Spine A sharp-pointed woody structure, commonly a modified branch or stipule.

Spur A term used to describe a short shoot-like structure bearing both flowers and leaves.

Spur-type A type of woody plant where the flowers are not simply derived from spurs, but the branches are inclined to produce an abundance of spurs rather than secondary branches.

Staminate The term used to describe a unisexual flower bearing only stamens (a male flower).

Staminode: A modified and non-functioning stamen usually resembling a petal.

Stellate Star-shaped.

Stipulate With stipules.

Stipule Small leaf-like structures found at the bases of leaf stalks of some plants. Stipules generally occur in pairs.

Strigose Hairy surface with all the hairs facing in one direction.

Stolon A horizontal stem that creeps along the ground, rooting at nodes and producing new plants.

Subshrub a plant intermediate between an herbaceous perennial and a shrub. The base is always woody.

Tendril A portion of a stem or leaf modified for climbing. Usually tendrils are slender and coiled. Some have adhesive tips.

Tomentose Wooly with matted hairs.

Trifoliate A compound leaf with three leaflets per leaf.

Truncated Ending abruptly as if cut off.

Umbel A flower cluster with numerous longated pedicels all emerging from one point, like the ribs of an umbrella.

Verrucose Warty.

WCSH The Western Canadian Society for Horticulture.

Whorl Three or more organs arranged in a circle.

Cross-Referencing Index

English Name	Latin Name	Starting on Page
Alder	Alnus	42
Almond	Prunus	120
Anemone	Anemone	44
Arborvitae	Thuja	159
Ash	Fraxinus	75
Asperula	Galium	77
Barberry	Berberis	46
Barvenok	Vinca	168
Basswood	Tilia	161
Bearberry	Arctostaphylos	45
Birch	Betula	49
Bittersweet	Celastrus	54
Broom	Cytisus	66
Buckeye	Aesculus	40
Buckthorn	Rhamnus	131
Buffaloberry	Shepherdia	146
Burningbush	Euonymus	70
Bush-Cranberry	Viburnum	164
Butternut	Juglans	83
Caragana	Caragana	52
Cherry	Prunus	120
Chokecherry	Prunus	120
Cinquefoil	Potentilla	118
Clematis	Clematis	55
Cliff-Green	Paxistima	100
Corktree	Phellodendron	101
Cotoneaster	Cotoneaster	62
Crab Apple	Malus	95
Crown Vetch	Coronilla	61
Currant	Ribes	134
Daphne	Daphne	68
Dogwood	Cornus	58
Douglas-Fir	Pseudotsuga	127
Duchesnea	Duchesnea	69
Elder	Sambucus	144
Elm	Ulmus	163
False-Spirea	Sorbaria	147
Fescue	Festuca	74
Fir	Abies	33
Fleeceflower	Polygonum	113
Forsythia	Forsythia	74
Garlandflower	Daphne	68
Goatsfoot	Aegopodium	40
Goldenbells	Forsythia	74
Gooseberry	Ribes	134
Goutweed	Aegopodium	40
Grape	Vitis	169
Groundcover Phlox	Phlox	103
Hackberry	Celtis	54
Hawthorn	Crataegus	65
Hazelnut	Corylus	61
Honeysuckle	Lonicera	89
Horse Chestnut	Aesculus	40
Hydrangea	Hydrangea	81

Woody Ornamentals

English Name	Latin Name	Starting on Page
Ivy	*Hedera*	80
Juniper	*Juniperus*	84
Kinnikinik	*Arctostaphylos*	45
Larch	*Larix*	88
Lilac	*Syringa*	153
Linden	*Tilia*	161
Maackia	*Maackia*	93
Maidenhair Tree	*Ginkgo*	79
Maple	*Acer*	35
Mockorange	*Philadelphus*	101
Mountain-Ash	*Sorbus*	148
Ninebark	*Physocarpus*	103
Oak	*Quercus*	129
Oleaster	*Elaeagnus*	69
Ononis	*Ononis*	97
Oregon-Boxwood	*Paxistima*	100
Oregon-Grape	*Mahonia*	94
Pear	*Pyrus*	128
Peashrub	*Caragana*	52
Periwinkle	*Vinca*	168
Phlox	*Phlox*	103
Pine	*Pinus*	108
Plum	*Prunus*	120
Poplar	*Populus*	113
Potentilla	*Potentilla*	118
Pricklyspine	*Acanthopanax*	35
Prinsepia	*Prinsepia*	119
Pussytoes	*Antennaria*	44
Rose	*Rosa*	135
Russian Sandthorn	*Hippophae*	81
Salt-tree	*Halimodendron*	79
Sandthorn	*Hippophae*	81
Saskatoon	*Amelanchier*	43
Saxifrage	*Bergenia*	48
Sea-Buckthorn	*Hippophae*	81
Shrub Rose	*Rosa*	135
Spindletree	*Euonymus*	70
Spirea	*Spiraea*	150
Spruce	*Picea*	104
Spurge	*Pachysandra*	98
Stonecrop	*Sedum*	145
Strawberrybush	*Euonymus*	70
Sumac	*Rhus*	132
Tamarack	*Larix*	88
Tamarisk	*Tamarix*	158
Trefoil	*Lotus*	93
Vetch	*Coronilla*	61
Viburnum	*Viburnum*	164
Virgin's-Bower	*Clematis*	55
Virginia Creeper	*Parthenocissus*	99
Walnut	*Juglans*	83
White Cedar	*Thuja*	159
Willow	*Salix*	140
Woadwaxen	*Genista*	78
Yew	*Taxus*	159